Organic Principles and Practices Handbook Series
A Project of the Northeast Organic Farming Association

Growing Healthy Vegetable Crops

Working with Nature to Control Diseases and Pests Organically

Revised and Updated

BRIAN CALDWELL

Illustrated by Jocelyn Langer

CHELSEA GREEN PUBLIS[illegible]
WHITE RIVER JUNCTION, VE[illegible]MONT

Originally published in 2004 as *Vegetable Crop Health: Helping Nature Control Diseases & Pests Organically.*

Editorial Coordinator: Makenna Goodman
Project Manager: Bill Bokermann
Copy Editor: Cannon Labrie
Proofreader: Helen Walden
Indexer: Peggy Holloway
Designer: Peter Holm, Sterling Hill Productions

Printed in the United States of America
First Chelsea Green revised printing March, 2011
10 9 8 7 6 5 4 3 2 1   11 12 13 14

Chelsea Green Publishing is committed to preserving ancient forests and natural resources. We elected to print this title on 30-percent postconsumer recycled paper, processed chlorine-free. As a result, for this printing, we have saved:

4 Trees (40' tall and 6-8" diameter)
1 Million BTUs of Total Energy
378 Pounds of Greenhouse Gases
1,822 Gallons of Wastewater
111 Pounds of Solid Waste

Chelsea Green Publishing made this paper choice because we and our printer, Thomson-Shore, Inc., are members of the Green Press Initiative, a nonprofit program dedicated to supporting authors, publishers, and suppliers in their efforts to reduce their use of fiber obtained from endangered forests. For more information, visit: www.greenpressinitiative.org.

Environmental impact estimates were made using the Environmental Defense Paper Calculator. For more information visit: www.papercalculator.org.

Our Commitment to Green Publishing

Chelsea Green sees publishing as a tool for cultural change and ecological stewardship. We strive to align our book manufacturing practices with our editorial mission and to reduce the impact of our business enterprise in the environment. We print our books and catalogs on chlorine-free recycled paper, using vegetable-based inks whenever possible. This book may cost slightly more because we use recycled paper, and we hope you'll agree that it's worth it. Chelsea Green is a member of the Green Press Initiative (www.greenpressinitiative.org), a nonprofit coalition of publishers, manufacturers, and authors working to protect the world's endangered forests and conserve natural resources. *Growing Healthy Vegetable Crops* was printed on Joy White, a 30-percent postconsumer recycled paper supplied by Thomson-Shore.

Library of Congress Cataloging-in-Publication Data
Caldwell, Brian Alan.
Growing healthy vegetable crops : working with nature to control diseases and pests organically / Brian Caldwell ; illustrated by Jocelyn Langer. -- Updated and rev.
p. cm. -- (Organic principles and practices handbook series)
"A Project of the Northeast Organic Farming Association."
Previous edition published under the title Vegetable crop health: helping nature control diseases & pests organically. Barre, MA : NOFA Interstate Council, c2004.
Includes bibliographical references and index.
ISBN 978-1-60358-349-7
1. Vegetables--Diseases and pests. 2. Vegetables--Diseases and pests--Control. 3. Agricultural pests. 4. Agricultural pests--Control. I. Northeast Organic Farming Association. II. Title. III. Series: Organic principles and practices handbook series.

SB608.V4C35 2011
635'.049--dc22

2010053672

Chelsea Green Publishing Company
Post Office Box 428
White River Junction, VT 05001
(802) 295-6300
www.chelseagreen.com

Best Practices for Farmers and Gardeners

The NOFA handbook series is designed to give a comprehensive view of key farming practices from the organic perspective. The content is geared to serious farmers, gardeners, and homesteaders and those looking to make the transition to organic practices.

Many readers may have arrived at their own best methods to suit their situations of place and pocketbook. These handbooks may help practitioners review and reconsider their concepts and practices in light of holistic biological realities, classic works, and recent research.

Organic agriculture has deep roots and a complex paradigm that stands in bold contrast to the industrialized conventional agriculture that is dominant today. It's critical that organic farming get a fair hearing in the public arena—and that farmers have access not only to the real dirt on organic methods and practices but also to the concepts behind them.

About This Series

The Northeast Organic Farming Association (NOFA) is one of the oldest organic agriculture organizations in the country, dedicated to organic food production and a safer, healthier environment. NOFA has independent chapters in Connecticut, Massachusetts, New Hampshire, New Jersey, New York, Rhode Island, and Vermont.

This handbook series began with a gift to NOFA/Mass and continues under the NOFA Interstate Council with support from NOFA/Mass and a generous grant from Sustainable Agriculture Research and Education (SARE). The project has utilized the expertise of NOFA members and other organic farmers and educators in the Northeast as writers and reviewers. Help also came from the Pennsylvania Association for Sustainable Agriculture and from the Maine Organic Farmers and Gardeners Association.

Jocelyn Langer illustrated the series, and Jonathan von Ranson edited it and coordinated the project. The Manuals Project Committee included Bill Duesing, Steve Gilman, Elizabeth Henderson, Julie Rawson, and Jonathan von Ranson. The committee thanks SARE and the wonderful farmers and educators whose willing commitment it represents.

TO MY FAMILY

This book is offered as a practical guide to effective organic-vegetable pest management for northeast growers, based on the experience and understanding of farmers and researchers.

CONTENTS

one

Introduction

Natural Models

What do we see when we look at a relatively undisturbed area, for instance, a forest or old field meadow? At first glance, we see a diverse mixture of primarily healthy plants. We see very little bare soil, no harvest removal, minimal nutrient deficiency, plants of different heights, and intense competition for light. Looking a little closer, we see some pest damage—galls on the goldenrod and a few holes in leaves. A great diversity of herbivores and predators are present, but at relatively low levels. Plant productivity is high. Many ecological niches are present beneath the plant canopy, fueled by decomposition.

Not so much is known about what is going on belowground. Roots explore the whole soil profile. Compaction is minimal. Aeration is as good

Natural models: wild field ecosystem (weeds with spiderweb).

as it can be, with hummocks creating zones of good drainage even on wet sites. The species composition of the soil community is very diverse, and differs from that of agricultural soil. Because of lack of tillage, fungi are better established. Soil organic matter is slowly increasing to high levels. Nutrient leaching is low.

We can try to gain insights from these natural examples, to help us better manage our crop ecosystems. But a farm field is profoundly different. In agriculture, we need to harvest crops and remove their organic matter from the field; we often need to till the soil, destroying a habitat above- and belowground, to establish our edible crop plants. Many crop plants are derived from early-succession plants that can take advantage of disturbed ground to grow and reproduce rapidly. Their pests are those insects that are able to locate their hosts within diverse landscapes, and also can reproduce rapidly once hosts are found. Agricultural systems are inherently different from less-disturbed natural models.

Human Requirements for Food Production

While there are many kinds of edible plants for humans in the wild, they are present in far lower concentration than in a garden or vegetable field. We require that our crops yield very heavily, compared to the edible offerings of the wild. A farm field offers up a niche of high-energy, easy-to-find, easy-to-digest abundance. Furthermore, because of centuries of human selection and monocultural planting practices, the crop variety found in the farm field may lack resistance to some pests. Not surprisingly, if we create huge niches, filled with concentrated energy and nutrients, then other creatures will try—and even evolve—to utilize them. In this manual, we will call this easy target the "Big Niche."

What Is a Pest?

The Big Niche is a golden opportunity for all manner of creatures. From their point of view, they are doing what comes naturally.

When they damage a crop or compete with us humans, we call them

Through the Insect's Eyes

An insect perhaps lives on a wild plant related to cabbage. Man comes along and cultivates whole fields of cabbages. Picture the huge joy of that insect when instead of hunting among crowds of other plants for its particular food it finds masses spread out temptingly in a manner beyond its wildest dreams. Picture the joy with which that insect hurries off to take a wife and bring into the world as many generations as possible in the limited time at its disposal, so that they will all live a life of ease and luxury free from gnawing anxieties. Wouldn't insects be fools if they missed such a glorious opportunity! And insects are not fools. (Cheesman 1952)

pests. Any type of organism is eligible for this term—insect, mite, fungus, bacterium, nematode, virus, mollusk, mammal.

We may become suspicious of any critters that we find in our fields. But in order truly to be a pest, an interloper must cause noticeable loss. Many organisms may live on or around our crops without competing with us at all. They just happen to be there, or maybe they are even predators hunting for potential pests. They may be performing essential ecosystem functions such as pollinating, eating detritus, or cycling nutrients. A key part of ecological pest management is to protect these non-pests, many of whom are doing good things!

In the last century, it has become all too easy to wage chemical war against pests who take advantage of the big niches we create. Organic farmers have chosen a different approach, avoiding the use of almost all pesticides. Along with that choice came the realization that the soil needed much more care and attention than it was getting. Over the past fifty years, organic vegetable growers have found ways to produce bountiful, high-quality crops on their farms using skill, science, and persistence.

This manual lays out pest-control methods pioneered by commercial organic vegetable growers in the northeastern United States. The ideas in it should be useful to any vegetable grower of any scale or persuasion.

Pests that will be covered in this manual include insects, other arthropods, and mammals that attack plants or infest their edible portions, as well as pathogens and other organisms that cause disease-like symptoms (pathogenic nematodes, etc.), ruining or reducing yields. Some symptoms, like nutrient deficiencies, that look like pest damage will also be discussed briefly in the pest ID section.

Weeds are pests, but are not covered here. See *Organic Soil-Fertility and Weed Management* by Steve Gilman in this series of NOFA manuals.

two

Basic Concepts: Clarifying and Focusing Our Thinking

Where do pests come from? To the inexperienced farmer or gardener, it appears that pests "come out of nowhere" to attack the crop. Of course, the pest organism, whether animal or pathogen, has to come from somewhere. It may have been lurking in low numbers in and around the fields for many years, until finally conditions were right for an outbreak. Pests tend to be organisms that evolved in disturbed environments and can rapidly proliferate when they find good conditions. (Aphids are prime examples. They can explode in numbers if predators are not present to keep their numbers down.) A crop field is a disturbed environment, much simplified above- and belowground compared to more natural areas nearby. Furthermore, most of our crop plants are not growing in their native areas, and at least half of our vegetable pests were introduced from other continents. The natural enemies that have co-evolved with these pests may not be present.

It is critical to learn the life history of a pest that is causing trouble. This involves where, when, and how many times per season the pest organism—microbe or animal—reproduces, and under what conditions; how and where it lives through the winter; and how and when it disperses to find the crop. Weak spots in the pest's life history can often be targeted to manage the pest without destructive effects on other species.

If crop pests are found in and around the field in high numbers, then *pest pressure* is said to be high. It is a key practice in organic farming to take steps to reduce overall pest pressure on the farm. This involves crop rotation, sanitation, and provision of habitat for natural pest enemies.

On a deeper level, "where do pests come from" can mean, "what *causes* pests to arise?" If we can answer this question, we can approach the root cause of pest problems.

For disease to occur, plant pathologists identify three conditions, referred to as the plant-disease triangle (Agrios 1988). These are (1) the

An Insect's Odds

An overwintered or newly hatched insect pest faces intimidating odds in finding its host plant. First of all, it is lucky to have made it through the winter. Many don't. In some years, nearly all of a pest species in a given locality may be killed by weather extremes.

Perhaps the farmer has rotated the crop from its previous location. Our pest must find the new planting, perhaps farther from its overwintering site. If the farmer delays planting the crop for long enough, the pest may starve to death looking for a host.

It may have to fly over a creek or a wooded area to get there, braving predators—birds, bats, dragonflies, other insects—and (ugh) even parasites the whole way! Life is tough. Once it gets to the field, the ever-scheming farmer may have intercropped the planting, making it darned difficult to find the right plant. Finally, the plant may try to defend itself, calling more predators and parasites, secreting yucky goo, and producing toxins! Life is tough—*but pests are good at what they do.*

susceptibility of the host; (2) the proper *environmental conditions* (often temperature, humidity, etc.); and (3) the *presence* and virulence of the pathogen. The level of disease can be thought of as proportional to the product of all three conditions. If any condition is absent, the pathogen is stymied and cannot cause disease.

This model is true enough for disease on an individual plant, but we need to enlarge it to cover all types of pests, in a field situation. We can do this by expanding the idea of environmental conditions beyond weather to include ecosystem factors such as the presence of natural enemies and competitors plus their ease of access to the crop. Then these same three conditions apply to all types of pests, over the whole farm.

In this manual, let's call the susceptibility of a plant *host susceptibility.*

Let's expand the environmental conditions to include *ecosystem factors.* The local population and virulence of the pest will be *pest pressure.*

The three legs of the pest triangle interact with each other. For instance, if ecosystem factors are non-conducive and the crop plant is not very

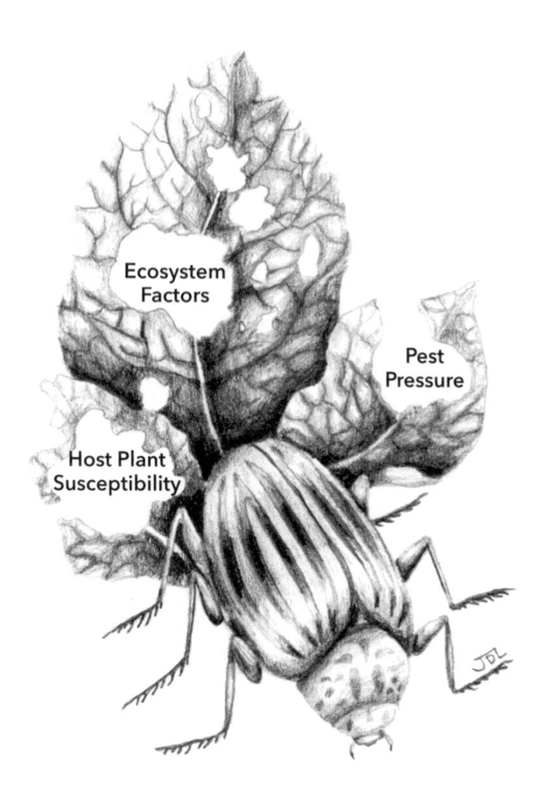

The plant-pest triangle shows the three factors involved in pest damage: susceptibility of host, environmental conditions, and presence and pressure of pests.

susceptible, then pest reproduction and survival are reduced, lowering pest pressure for the next generation.

When host susceptibility is high *and* ecosystem factors are conducive to the pest, we have a Big Niche. This is the situation that develops for most farm crops today. If pests are present, the only management choice is to kill as many of them as possible, since the crop is a sitting duck.

Host Susceptibility

How do plants resist pests? Probably 99 percent of potential pest species are resisted by a given plant because the pest organism simply cannot use the crop as a food source, or home. In other words, the organism cannot utilize the niche that the crop provides. In human terms, we can't eat grass, because we just can't digest it—even though it is relatively high in nutrients for animals with four stomachs! Similarly, very few fungi can utilize a crop of, say, melons. But those that can are specifically adapted and glom right on to the crop if given the chance. So—how do plants resist the pests that are adapted to utilize them?

Plant-Positive versus Pest-Negative

According to Eliot Coleman (1995), "well-grown plants are insusceptible to pests . . . plants only become susceptible to pest attack when they are stressed by inadequate growing conditions. Thus, [wise gardeners] see pests not as enemies of plants, but as helpful indicators of cultural practices that need to be improved." Coleman coined the term *plant-positive*, as opposed to *pest-negative*, for this approach. The concept is not new.

The 1966 Rodale book *The Organic Way to Plant Protection* stated that "a healthy plant, grown on a balanced fertile soil, can and will resist most insect pests" (p. 31). In 1973, entomologist Cynthia Westcott (1973) was dubious about this claim: "Advocates of organic gardening claim that plants grown without chemical fertilizer but with proper organic culture are resistant to pests. Vigorous plants may more readily survive the devastating effects of pests, but on the other hand some pests, such as aphids, are attracted to lush, succulent tissue. . . . [T]his field is relatively unexplored." In fact, since Dr. Westcott's time, there has been considerable research showing that plants in optimal soil conditions can resist pests—but recall that this is only one side of the pest triangle.

Soils and Pests

There are many studies showing that for some pests, aspects of plant nutrition can make plants more or less susceptible.

It is well accepted that overly high levels of nitrogen fertilization, resulting in the accumulation of nitrate or free amino acids in plant leaves, can increase outbreaks of pest aphids and mites. However, low levels of nutrients can also create problems. The key seems to be a long-term strategy of fostering balanced soil fertility and high soil biological activity (Phelan et al. 1996).

Proper plant nutrition results in slower pest feeding and reproduction, and may prevent a mite or aphid explosion. However, for full control, it is also necessary to have enough of the ecosystem control agents for these pests—natural enemies such as lady beetles, predatory mites, lacewings, etc. The reduced host susceptibility and ecosystem factors work together.

Now we get into some recent groundbreaking research on a major sweet-corn pest, the European corn borer, *Ostrinia nubilalis* (ECB). In greenhouse trials, ECB females preferred to lay eggs on corn plants grown in soils from chemically fertilized fields, versus those grown in paired soils from long-term organic fields (Phelan et al. 1995). This finding clearly shows that the management of the soil can affect the attractiveness to pests of plants grown in it. The greenhouse data are backed up by the authors' statement that they measured about six times as many first-generation ECBs and four times as many corn rootworm eggs in second-year corn on a conventional versus a paired organic farm.

In a second study (Phelan et al. 1996), the best greenhouse results were from plants that were grown with compost fertilizer added to organic soil that had previously been treated for over twenty-five years with uncomposted manure and legume plow-downs in a red clover/grass–corn–soybeans–small grain rotation. However, the same compost fertilizer applied to the conventionally fertilized soil resulted in the highest number of eggs laid; even more than the same soil with ammonium nitrate or no fertilizer. Clearly, it may take more than one year of organic practice to achieve desired results. This underlines the problems that can come up in the transition of fields from conventional to organic management.

The authors speculate that the difference in ECB egg-laying behavior may arise from different mineral balances in plants grown on the two soils. However, it is also possible that some other unmeasured factor(s) are responsible. We need to keep in mind that greenhouse conditions are very different from those in the field. However, under these conditions, the study clearly demonstrated reduced host susceptibility because of lowered attractiveness to egg-laying pests.

Soil Health

Besides soil nutrient levels, soil health is very important to overall crop performance. Soil health refers to the quality of the soil environment for the crop root system. Lack of compaction and good drainage, plus a healthy soil biology, are the key aspects of soil health.

Cotton Studies

As described by Dr. Joe Lewis of the USDA's Insect Biology and Population Management Research Lab in Tifton, Georgia, the cotton plant brings a powerful array of natural defenses against pests. When attacked by insects, the plants release "volatile chemical cues that attract predators and parasitoids that, in turn, attack the herbivores" (Lewis et al. 1997). A wilder strain of cotton releases around ten times as much of these volatiles as do typical modern commercial lines. Furthermore, cotton plants also provide extra-floral nectaries—nectar-secreting organs in addition to their flowers—to feed and encourage parasitoids. The cotton plant brings with it these cooperative weapons against herbivore pests—but they only help in an ecosystem that also provides habitat for beneficials. It is important to note two more things. First, the extra-floral nectaries also provide food for pest moths. This makes sense, since a reliable, though limited, population of pests is necessary to maintain predators. Second, the plant doesn't release its volatile cues until actually attacked. We can infer that production of the volatiles requires energy, so the plant will not produce them unless they are needed; also, it would be detrimental to the parasitoids if they were attracted to plants in the absence of their pest prey.

Do other crop plants perform similar wonders? Undoubtedly. Can we select and breed crops for these abilities? What an opportunity for some very worthwhile research!

Induced Resistance

When plants are attacked by herbivores or pathogens, many are able to produce specific defensive compounds in their tissues that reduce further damage. These defensive compounds can spread throughout the plant, and may even be active against new attacks by completely different pests. This response is called "induced resistance" or "systemic acquired resistance."

Such resistance can be stimulated by spraying plants with weak pathogens, special proteins, and other chemicals. Some composts and compost-treated soils can induce resistance to pests that attack the aboveground parts of the plant. While this effect has been clearly demonstrated, getting consistent results in the field with products formulated to induce resistance has been difficult. One such product available to organic farmers is Trichoderma strain T-22, sold as PlantShield®.

The strength of the resistance effect in the plant depends both on the material used to induce resistance and the genetic ability of the crop to respond.

Genetic Resistance and Plant Breeding

Plants can resist pests by expressing specific traits governed by one or a few genes. For instance, resistance to the tomato disease verticillium wilt, among many others, may occur this way. The typical method used by modern plant breeders to endow a crop plant with resistance to a given pest is to isolate the resistance-carrying genes, breeding them into a few individuals. These are then backcrossed with the non-resistant crop until a commercially acceptable, resistant variety is obtained. This sort of resistance can be very specific to a given disease or insect pest. Its ultimate expression is genetic engineering, an extremely reductionist approach that bypasses sexual reproduction entirely and ignores holistic approaches. Organic farmers reject GMO varieties. However, use of non-GMO pest-resistant varieties can be an important component of an organic pest-management program.

There is another type of resistance that is general. One might call it the overall vitality of a given plant. Though there are many differences, this general resistance of the plant corresponds roughly to the general immune response in animals. As in animals, the better the overall condition of the plant, the more resources it has available to use against a pathogen or herbivore. The response to pest attack requires energy and nutrients, as when a plant encountering fungal proteins responds by synthesizing a specific anti-fungal substance.

One strategy that disease organisms employ is to attempt a sneak attack, infecting the plant without triggering its general resistance. Sometimes a resistant variety merely has a more sensitive trigger for general resistance than most, so that it kicks in before the disease or pest can get established.

Evolving Pests

The pest organism itself is constantly evolving to be more effective at utilizing the Big Niche. Pests may have started out as introduced or native insects—or diseases, or nematodes, or mammals—that happened upon a farm crop they found they could use. Then, however, a process started in which the pest began propagating in or near agricultural operations and spread from region to region through many possible means. Eventually, though the human awareness may have been sudden, one had a major pest like the Colorado potato beetle or apple scab. The plum curculio, a major apple pest in the northeastern United States, is thought to have been adapted only to wild plums, other stone fruit, and perhaps hawthorn until the mid-1800s when it made the jump to apples.

In contrast to the "isolate and backcross" approach to plant breeding, indigenous peoples over thousands of years bred plants by mass selection—with tremendous results! This meant that a fairly large, diverse crop population was purposely maintained. Under various conditions and the keen eye of the farmer, strains were selected that resisted pests or filled other needs.

Such mass or horizontal selection breeding will prove increasingly valuable in holistic systems—as we relearn how to do it.

Itching for an Answer

It is clear from related personal experience, however, that health or vigor does not always leave one free of pests. For example, few of us could walk unprotected in swampy areas in midsummer and emerge without being attacked by mosquitoes. Does that mean that we're all in poor health? No—obviously, the ecosystem we are in profoundly influences the pest populations and pressures we encounter.

Ecosystem Factors

In this manual, both plant-positive approaches that *reduce host susceptibility by increasing plant health and genetic resistance* and pest-negative approaches that work by modifying ecosystem factors will be emphasized. Direct assaults against the pest—i.e., pesticide sprays, even with approved products—ideally are the last resort of the organic farmer, because they have so many off-target impacts. Remember, our goal is to reduce that Big Niche to a small one.

Effects of Cultural Practices

Besides altering host susceptibility, cultural practices may also affect ecosystem factors, in turn influencing pest pressure (see chapter 3). For

Cultural Practices That Benefit Pest Control

Do they decrease host susceptibility (HS), modify ecosystem factors (EF), or both?

Beneficials habitat	EF
Balanced ecosystem	EF
Row covers, hoophouse, etc.	EF
High-pressure water sprays	EF
Sanitation	EF
Organic mulch	EF
Crop rotation (avoidance)	Both; mostly EF
Foliar sprays	Both
Green manures	Both
Proper spacing, irrigation, and site	Both
Timing of crop season	Both
Good soil health	Both; mostly HS
Compost applications	Both; mostly HS
Fertility adjustment	HS
Proper variety	HS
Treatments to induce resistance	HS

instance, the practice of providing habitat for beneficials does nothing to improve crop growth: it is actually pest-negative, fostering natural enemies, making the ecosystem less hospitable to the pest, thereby reducing pest pressure. In a similar vein, high levels of soil organic matter may prevent some diseases not only because they provide better conditions for crop growth (reducing host susceptibility), but also because their inherent microbiological diversity (an ecosystem factor) prevents pest organisms from building up.

How Can an Ecosystem Keep Pests in Balance?

In addition to healthy soils and plants, the key aspects of a healthy agricultural ecosystem include:

- A good crop rotation.
- Proper variety selections, growing conditions, and timing.
- Undisturbed areas and planted areas that act as reservoirs and habitat for beneficial organisms (provision for birds and bats?).
- Crop sanitation.
- Diversity within the field.
- Feeding the soil ecosystem with carbon-rich inputs.

Good management of ecosystem factors will reduce pest pressure.

Flea beetle damage on young broccoli shoot.

Avoidance

Avoidance is a very important tactic in pest control. A pest can be dodged by timing (examples are planting wheat after the Hessian fly–free date; planting late to avoid Colorado potato beetle; planting all brassicas after July to avoid crucifer flea beetle; planting only early corn to avoid corn earworm), or by locating the crop where pests are not present in sufficient numbers to cause harm. For example, some fields may be infested with the clubroot disease organism, making production of cruciferous crops impossible there. If so, it will take eight years or more before a good crop of crucifers can be grown again in these fields. A less obvious example occurs with the diamondback moth. This is a major pest of brassicas in heavy production areas, but in my experience, on many isolated organic farms it is not a problem. The same thing is often true for bacterial diseases on tomatoes and peppers. In these cases, ecosystem factors plus extremely low on-farm pest pressure reinforce each other to keep pest pressure extremely low. The message here is to beware of importing any new pests to your farm!

Q: What is sanitation?

A: Weakening pest pressure by reducing or eliminating situations that can harbor pests. Sanitation includes such things as disinfecting greenhouse trays, pots, flats, and surfaces with an approved product or by exposing them to excessive heat during the off-season; tilling under crop debris before seeding a cover crop; using clean (pathogen-free) or organically-treated seed; in an extreme case—hosing down equipment to prevent spread of a soil-borne disease or weed seed from an infested field.

Once a disease or even an insect is introduced onto an organic farm where it was previously absent, it often becomes a fairly persistent problem. This points up the need for careful sanitation and selection of clean seeds and plants.

Field Complexity

The more complex a field is, the more niches there are for generalist predators. However, there are also more niches for generalist pests! Good farm design must account for these generalists (see chapter 4).

Complexity in the field includes diversity of crops, weeds, sod strips, mulches, and provision for crops and/or perches of differing heights within the field. Let's take these one by one.

Diversity of Crops

Crop diversity has been shown in many scientific studies to decrease pest damage. Pests have a harder time finding their crop. An example is Richard Root's (1973) classic study of collard intercropping. Flea beetle incidence and damage was decreased in tomato-intercropped collards compared to monocropped collard plantings. It is likely that the scent of the tomatoes masked that of the collards; also the mere physical presence of an obstructing plant makes it harder for flea beetles to invade a dispersed collard crop. In other cases, an intercrop species may provide nectar or habitat for beneficials, which then attack the pest. Intercropping systems can range from large alternating strips of eight rows or more, to companion planting of different species within the row.

Warning: Sometimes pest damage increases in intercropped situations. Generalist pests such as tarnished plant bugs, stinkbugs, viruses, voles, slugs, or even woodchucks may thrive in fields where their habitat and hosts are more prevalent *because* of diversity. Management of habitat near or in the fields by mowing or roguing can reduce problems with generalists. If you find one of these generalist pests seriously affecting a crop, try to identify its other hosts and eliminate them from the area if possible. Or, you may have to remove the affected crop from the mix.

Weeds in the Field

Weeds provide habitat and food sources for beneficials the same as interplanted crops do. Besides heeding the same warning above, it is crucial in profitable organic-crop production to *keep weeds from going to seed* as much as possible. And of course, too many weeds will reduce yields and quality just like other pests. Heavy weed populations also increase humidity around the crop, increasing disease. So cropland weeds can help with pest control but must be handled judiciously, since they are pests themselves. They may harbor pests and/or beneficials.

Sod Strips

Leaving strips of sod within the field is a form of intercropping. Since they are untilled, they encourage soil-dwelling species such as ground (carabid) beetles (Pickett and Bugg 1998). Carabid beetles eat weed seeds, small insects, larvae, and eggs that they find in the soil. They can regularly move at least ten feet into a tilled field from a sod strip. Sod strips are also very helpful in providing aisles for harvest or other field operations. However, voles in these strips can cause serious problems if the sod is mowed infrequently.

Mulches

Using mulch has many benefits. Mulches provide in-field habitat for ground beetles, spiders, and garter snakes. They also reduce weeds, conserve moisture, and can be used to modify temperatures to suit the crop. They increase soil quality. Finally, it seems that some pests such as Colorado potato beetles and flea beetles just don't like organic mulches (Ruutilla 1999). However, slugs and possibly secretive pests like squash bugs and stinkbugs may be increased in mulched crops.

Crops of Different Heights

Occasional tall trellised or caged crops like tomatoes can provide bird perches in the field. The effects of birds on vegetable pest control needs more study, but many gardeners and some farmers feel they are important.

Yield Loss

Intercropping systems using cash-crop species should be designed to result in little, if any, yield loss. In fact, if the crops are complementary in the way they use soil and light resources, their combined yield may actually be higher than if they were grown separately. This is called over-yielding. On the other hand, adding noncash cover crops to your field will likely reduce your overall cash-crop yields per acre. But check out your yields per dollar or per hour—they may go up!

Habitat around the Field

It is important that the area around the field serve as a reservoir for beneficial species. If you provide food sources for beneficials in the field, but the area around the field has none to offer, the effect on pests will be far less than if neighboring areas held a good supply. Hedgerow tree and shrub species and woodlands are particularly valuable sources of beneficials (Marino and Landis 1996, Altieri and Nicholls 1998). A well-mowed strip of sod of at least twenty-five feet between the vegetable field and any brushy area will help reduce damage by voles, woodchucks, and rabbits. Luckily, flying beneficials will still easily disperse into the vegetables.

Some weedy plants outside the field can host viruses. Overwintering broadleaf weeds such as pokeweed and burdock, etc. should be controlled by mowing.

Integrated Growing Systems

A reductionist's attempt to tease apart what is going on in an effective organic pest-management system will often be frustrated. Indeed, in the absence of hard science, the farmer uses art and skill to allow a set of mutually reinforcing strategies to manage pests. They include soil management intertwined with crop rotation and other cultural practices, intertwined with management of the surrounding landscape.

David Andow of the entomology department of the University of Minnesota and Kazumasa Hidaka of Hiroshima University (1989), in their study comparing *shizen* (natural) with conventional rice production in Japan, make the point that the two contrasting approaches are made up of distinct groups of practices that function together to form integrated management syndromes. The methodologies produce similar high yields. The *shizen* fertility and irrigation system shows plant growth that starts slower but is stronger by the end of the season, allowing for less disease and more support of beneficials populations; the conventional system uses rapid early growth and pesticide support to attain its yield and financial goals. For more on holistic farming management, see the accompanying volume in this series, *Whole-Farm Planning*, by Elizabeth Henderson and Karl North.

How Is Organic Pest Management Different from IPM?

Integrated pest management (IPM) has many of the same goals as organic agriculture—producing generous yields of high-quality food with as little adverse environmental impact as possible. However, there are key differences. Essentially, IPM starts with typical farm practices and tries to improve them. By contrast, organic farming rejects at least two basic conventional farm methods: chemical fertilization and chemical pest control. Organic farming necessarily requires a fundamental emphasis on cultural methods and soil and plant health to manage pests. It is a radical, "at the root," approach to farming.

In IPM systems, pest pressure and weather conditions are scouted frequently. If pests or conditions exceed a cost-benefit threshold, a pest-management intervention is made, usually a spray.

IPM thresholds are formulated by estimating how much monetary damage pests would cause to a given crop, and comparing that to the cost of the intervention. The thresholds used in a conventional system often don't apply on an organic farm. The costs of the crop and the intervention may be quite different than for a conventional farm. Also, the organic farmer may decide not to intervene simply because to do so would be too disruptive or dangerous. She may just swallow the loss and make management changes for next year, especially if the crop will return a decent profit anyway. Of course, the IPM farmer can do this too, but since the IPM system is built around the idea of a cost-benefit threshold, it is less likely.

Organic growers tend to get premium prices for their crops. This can affect the perception of pest threat in two different ways. Some growers might think that since the crop is worth more, a small amount of damage results in a significant loss of income; so intervention is warranted at a lower threshold level. Others may think, "I'm already making good money on this crop, I can afford to lose some to pests." This attitude results in tolerance of more damage before intervention.

In another example of how costs are handled differently, organic farmers make extensive use of row covers to shield crops from pests. Strict IPM methods find this approach far too expensive to recommend compared to an inexpensive pesticide spray.

Results from a Cornell vegetable systems trial support the idea that organic vegetable production methods are more environmentally sound than IPM methods; the ongoing question in this study is whether they are more or less profitable (Petzoldt, unpub.).

In theory, IPM embraces all sorts of pest-management options, including all the approaches mentioned in these chapters. In practice, most IPM growers, however, place their emphasis on pesticide management. Also in theory, organic growers use pesticides only "as a last resort." However, in some cases where an effective organic pesticide exists, such as copper products against tomato early blight, organic growers may become lax in their use of cultural controls and over-spray.

When organic farmers are unsuccessful in removing the reasons for a pest outbreak, IPM provides a framework for making intervention decisions that they can use. This includes understanding the pest life cycle, monitoring to determine when intervention is needed, and choosing materials that will be effective in reducing pest pressure while disrupting ecosystem balance as little as possible.

three

Practical Approaches

Crop Cultural Practices

Site and Soil

Plants and pests are heavily influenced by the world around them. For a plant, the soil and weather are primal factors. Our job as farmers is to give the crop the best conditions we can. Site conditions are a good place to start. Ensure that the site is appropriate for the crops you are growing; conversely, if it is inappropriate, don't grow the crop, or you will have to compensate for its shortcomings. Most vegetable crops require full sun, sufficient air movement to dry off quickly in the morning, enough water, and the appropriate temperature and day-length ranges. The importance of good soil drainage cannot be overemphasized. Sites can be modified, up to a point, by cutting trees, installing drainage tile, using raised beds, irrigation, and so on. Give the plants what they need.

A more subtle aspect of site conditions is *aspect*, meaning which direction a sloping field faces. South-facing fields warm up quickly and are good for heat-loving crops. In our latitude, according to the *1913 USDA Yearbook*, a 5-degree slope to the south shifts the microclimate 300 miles south in terms of sun and warmth. Alternatively, north-facing slopes are better for cool-weather crops grown in the heat of summer. Similarly, sandy or gravelly soils warm up earlier than heavier ones. Experienced farmers are well aware of which fields are best for which crops. That being said, the benefits of a diverse rotation are so great that those fields will not always be planted in the most favored crops.

Soil nutrient levels can be changed much more easily than texture or drainage patterns. Get your soil tested, and use sound organic practices including manures, composts, rock powders, and cover crops, to bring available nutrient levels into the "high" ranges. Do not over-apply any particular nutrient. Extra-high available soil phosphorous has negative effects on water quality, and is unlikely to help your crop.

Some outstanding organic growers caution not to allow soil-test magnesium levels, in particular, to get too high relative to calcium. In practical terms, if your soil test shows a magnesium base saturation level above 20 percent, use a high-calcium lime to raise pH if needed, or gypsum to supply more calcium without raising the pH. Vegetable growers are also warned to avoid raising potassium levels beyond 8 percent of base saturation. Some growers follow even stricter cation balancing rules. (See *Soil Cation Nutrient Balancing in Sustainable Agriculture: Missing Link or Red Herring?* by Mark Schonbeck of the Virginia Association of Biological Farmers for more information.)

A common pitfall is to apply too much nitrogen relative to the other nutrients. This is known to worsen mite, aphid, and other insect problems in many, but not all, cases (Scriber 1984). Some plant diseases are also encouraged by overly lush growth resulting from excessive nitrogen. Fire blight in apples and pears is a prominent example.

Soil health is also critical in producing a healthy, optimally grown plant. Good soil health means a good environment for plant roots, including high levels of active soil organic matter (relatively new residues) and freedom from compaction and excessive tillage. Some soils actually become disease suppressive because of the resilient, highly active biological community engendered under good organic management.

Irrigation

Using drip irrigation instead of overhead watering greatly reduces the amount of time that your crop leaves are wet. This will reduce almost all fungal and bacterial diseases. If you must use an overhead system, try to apply water in midday so that the leaves dry quickly. This is contrary to common irrigation advice, since more water is lost to evaporation, but short periods of leaf wetness in midday are less likely to foster fungal diseases than adding more hours to the long leaf-wetness period that occurs during the night. Midday watering can be also be used to cool a crop such as peppers at a stressful time.

Resistant Varieties

Varieties resistant to many pests are offered in seed catalogs. Trial them on your farm to make sure that they have the qualities you need. Obviously, growing a plant with strong innate resistance to a pest can be our first line of defense against a problem pest. Unfortunately, you will probably find that many resistant varieties do not meet your needs. More work in this area is needed.

The number of publicly supported plant breeders has declined over the past decades—a real policy mistake. Another mistake is that many of the university breeders who are still around, and many seed company breeders, are focusing on genetic engineering as their primary breeding tool. This extremely narrow approach to crop health is fraught with serious potential problems for the environment and the consumer.

Cover Crops–Vegetable Farm Work Horses

Cover crops can fit into almost any vegetable-crop rotation. They are grown at times when the cash crop doesn't occupy the soil, and can even be undersown into standing cash crops. Cover crops perform several major beneficial functions. They:

- cover and protect the soil;
- absorb and recycle nutrients, reducing leaching;
- compete with weeds;
- fix nitrogen—if legumes—which can be the most important annual nutrient input on a well-managed farm;
- interrupt pest life cycles;
- add active organic matter to the soil, feeding the soil biological community; and
- provide food and habitat for beneficial insects (and also some pests!).

The most common cover crop is winter rye. It will establish well when planted after most vegetables are harvested, and provide good growth the following spring before being tilled in. However, there are many more ways of using cover crops, and many more candidate species. Please refer to the following excellent references for more information: The *Northeast Cover Crop Handbook* by Marianne Sarrantonio (1994), and *Managing Cover Crops Profitably* by the Sustainable Agriculture Network, 3rd ed. (2007).

Diverse intercropping at West Haven Farm, Ithaca, New York. PHOTO BY BRIAN CALDWELL.

Intercropping

Research supports use of intercropping with the proper mix and management of species. For instance, corn and soybeans complement each other well. They share few pests, and corn benefits from extra sunshine next to the soybean strips, while the beans are able to tolerate the corn competition without much yield loss.

Vegetable growers can mix strips or beds of various compatible vegetables in a field. For instance, one could plant successive herb crops—rows of arugula, cilantro, and dill—interspersed with strips of lettuce plantings. This sort of mixing it up, as opposed to single-crop blocks, can slow the spread of disease and make it harder for pest insects to find your crops. (Note that management of these crops, such as harvest, weed control, irrigation, etc., can be more complicated). I have seen many small-scale diversified organic vegetable farms with this sort of mixed-field approach. And while I don't have data to back it up, I believe that they are getting significant pest control benefits from this practice.

Experiment on Your Farm with Strip Cropping

Strips could be various widths from single rows to eight or twelve rows wide. Above this width, the diversity effect is probably lost, and the strips begin to function more like narrow monocrops.

Gardeners' lists of companion crops will help a farmer experiment with strip-cropping combinations (Cunningham 1998).

Insectary Plants

Plants can be good food sources or habitat for beneficial natural enemies of pests in several ways. They can provide a direct food source (i.e., nectar or pollen) for the beneficials; they can host prey (like aphids) that are a food source for the beneficials; and they can provide shelter for the beneficials.

The following plants are commonly mentioned as good insectary species. This is a brief coverage of this potentially vast subject. We need more research done in this area!

- Corn is a great crop for rearing beneficial insects, especially lady beetles, lacewings, and *Orius* bugs. It makes copious pollen, and is often a source of aphids for beneficials to eat.
- Composites—dandelion, aster, cosmos, black-eyed Susan, coreopsis, marigold, yarrow, tansy. These and other small-flowered plants listed below provide nectar for beneficial wasps and syrphid flies.
- Umbellifers—dill, caraway, fennel, angelica, lovage, parsley, Queen Anne's lace, coriander, bishop's weed (*Ammi majus*).
- Mints—spearmint, oregano, basil, etc.
- Crucifers—alyssum (also hosts crucifer flea beetles).
- Buckwheat.
- Clovers harbor many beneficials, but also aphids and tarnished plant bugs. This can be useful if the aphids provide food for beneficials during times when cash crops are absent. Also fava beans, black locust, cowpea, vetch, sweet clover, alfalfa. See the tarnished plant bug discussion under lettuce in chapter 7.
- Untilled sod is a refuge for ground beetles and mycorrhizal fungi. However, grower Jim Crawford of western Pennsylvania cautions about fostering Japanese beetles in sod. This may be a moot point if there are lawns, pastures, or hayfields near your farm.

All sorts of questions arise when working with intercropping, beneficial insect habitat, and farm design. When do the beneficials stay on their insectary plants? When do they move into crops? Which beneficials are most important? How much land area needs to be devoted to them? And on and on. An excellent book on this topic is *Enhancing Biological Control* by Charles Pickett and Robert Bugg (1998). The book is a classic, chock full of information from many scientific studies.

Mechanical Barriers

Row covers are effective at screening out some pests when properly used. Many organic growers are successful using row covers against flea beetles, cucumber beetles, cabbage root maggots, and other difficult pests. Row covers also allow for extra-early and extra-late harvests. Row covers only work when a mobile stage of the pest needs to find and infest the crop.

If the pest is already present (for instance, in the larval stage) when the row cover is applied, it will have no benefit and may even increase damage. However, in such cases growers have reported successful control by spraying right through the covers.

The cost of row covers is considerable. A lightweight material (0.5–0.6 oz/sq. yd.) if used for two seasons may cost around $300 per acre per year,

Row cover at Food Bank Farm, Hadley, Massachusetts. PHOTO BY BRIAN CALDWELL.

including shipment (2003 figures). A heavier material (0.9–1.25 oz./sq. yd.) that is also useful for frost protection may cost $500 per year, even if it can be used for four seasons. Since the same piece of row cover might be used for several different crops during a given year, these numbers may be reduced accordingly. Nevertheless, it is hard to justify the use of row covers on lower-value crops.

Sealing the edges of the row cover effectively, while still being able to remove it and replace it for weed-control operations, is another tricky and costly challenge. Black plastic bags filled with soil and placed at 6- to 15-foot intervals along the edge are proving to be an efficient alternative to soil in holding down edges. These are available from suppliers. Stakes have the disadvantage of tearing and weakening the edges of the cover.

Storage of covers in a dry, dark place when not in use helps to prolong their life over several years.

Trap Cropping

This used to be in the category of "a good idea, but needs to be proven." But recent work by Jude Boucher (2003) of the University of Connecticut Extension System and John Mishanec (2003) of Cornell Cooperative Extension presented at the New York State Vegetable Conference show that the idea is coming of age. John described a system that is effective against Colorado potato beetle (CPB) on farms that are not in heavy potato production areas.

First, the farm is divided into two halves, and all the potatoes, tomatoes, and eggplants are grown on one of the halves. In the next year, these crops are shifted to the other half of the farm. In a field between the two halves, a trap crop can be planted that will intercept the CPBs as they travel in search of a new crop. A row of a potato variety like Superior, which grows well in cool weather, is planted early in the season in a line between the two halves of the farm. It will strongly attract the CPB adults. The insects could be handpicked, sprayed, or the whole trap crop could be flamed when the fourth instar larvae appear (these larvae are the ones that are as big as the adults).

Does it work? John quotes IPM grower John Gade of Albany County: "It works. Next year we will definitely use trap cropping again because it took so little time and it saved us the hassle of treating the whole field."

Jude Boucher reported groundbreaking work to make perimeter trap cropping commercially viable. This approach has proved effective against diamondback moths (DBM) in on-farm trials when two rows of collards surrounded the main crop of cabbage. Though the fields were sprayed for cabbage loopers and imported cabbageworm, DBM never reached the threshold for intervention, and total spray costs were reduced by over $50 per acre. When the collard plants were unsprayed, a naturally occurring wasp, *Diadegma insulare*, parasitized 70 to 80 percent of the DBM on them. So a further benefit of the system was in the rearing of many beneficials.

This trap cropping design was pest-specific to DBM and did not affect cabbage loopers or imported cabbageworms. However, of the three, DBM is the most difficult to control, especially with Bt products.

Surrounding bell pepper plots with cherry peppers reduced pepper maggot damage from 15 percent to less than 2 percent.

Finally, when experimental yellow-squash fields were surrounded by blue Hubbard squash, 94 percent of all the striped cucumber beetles (SCB) were found on the Hubbard plants. Additionally, in trials where the perimeter Hubbard plants were sprayed, squash-vine-borer damage on the interior yellow squash was reduced by 88 percent. Boucher reports that six commercial growers successfully used this method in 2002. It is important that the trap-crop plants form a full perimeter around the protected crop and that the perimeter be in place as soon as or before the crop comes up.

The chart in chapter 7 showing relative striped cucumber beetle preference for different cucurbit varieties can be used to select trap crops. Note that there must be a strong difference in preference for the trap crop to work. In experimental work done in 2001, dark green zucchini was chosen as the trap-crop variety because it is highly preferred by SCB, it has a bush growth habit, and it takes up little room in the field. In one observation on a row of mixed dark green zucchini and New England pie pumpkin, all SCB were found on the trap crop. This effect did not extend to the next row 12 feet away, however, where the pumpkins had normal levels of SCB (Caldwell, unpub.)

Clearly trap cropping has much potential. We are just figuring out how to do it. Organic growers can make use of spraying or flaming of infested trap crops, particularly on cool mornings when pests are relatively immobile. Flaming eliminates the trap crop, so it should only be used if the trapping effect has accomplished its goal, or if there is another trap crop in place.

Sticky Tape

Sticky tape can be used with trap crops or over cash crops that are treated with repellants. Chris Petersheim of Paradise Farm in Paradise, Pennsylvania, uses this system.

He uses yellow sticky tape called Insect-A-Peel from Arbico (www.arbico.com). Chris says it is effective against cabbage root maggot flies when placed above crucifers that have been sprayed with a repellant such as EM-5 (Effective Microorganisms, see http://www.emtrading.com/em/htmlpapers/emcucumbers.html) or Surround®. It can be used the same way against leaf miners in beets and chard. Chris also uses it to monitor for the presence of some pests. The key advantage of this device is that the sticky yellow surface can be quickly renewed after it gets covered with dirt or insects. It is a nonspecific control, and will attract and kill beneficial insects as well as pests.

Rescue Treatments–Allowed Materials and Their Efficacy

Rescue treatments are sprays used when cultural practices, rotation, and other methods have not provided sufficient pest control. The spray materials are subject to approval by several layers of government before they can be used in certified organic production. Except for some specifically exempt materials (see the appendix), they must be approved as pesticide products by the EPA and legally labeled for the crop and pest. They must also be registered in whatever state they are used. Finally, *any product used* must be allowed under the USDA National Organic Program (NOP).

Currently, the Organic Materials Review Institute (OMRI; www.omri.org) maintains lists of products that meet NOP requirements. Some new products are also now labeled for use under the NOP by the EPA. Many products currently neither on the OMRI list nor so labeled would also be approvable under the NOP, but *all* the product ingredients must be known for that determination to be made. Since manufacturers are reluctant to divulge their proprietary secrets to the public, checking the OMRI list or NOP-approved label is the safest way to make sure that a given brand-name product is approved for use in certified organic production. In fact, give your certifier a call to be sure. Never rely only on the word of a salesperson or friend.

The *Resource Guide for Organic Pest and Disease Management* (2005) is a valuable tool for selecting rescue treatments. It is available for sale or at http://www.nysaes.cornell.edu/pp/resourceguide/.

Simple Materials

Refined oils smother pests and are usually effective only on soft-bodied pests such as aphids and mites and their eggs. I have never seen definitive data from trials with oils on leafhoppers, though it seems likely that nymphs would be affected. Oils also reduce powdery mildew. They have little residual effect. Be aware that oils, soaps, bicarbonates, sulfur, and copper products may all have plant-toxic effects if used under the wrong conditions. See the *Resource Guide* mentioned above for details.

Soaps destroy the membranes of aphids. They also have minimal residual effect.

Bicarbonates are mild fungicides, effective against powdery mildew and some other fungi.

Water sprays or heavy rains can destroy some pests such as mites.

Sulfur is a moderately strong fungicide and miticide. It has some residual activity, but is not systemic and is not able to kill fungal infections once they have penetrated into plant tissue. Sulfur is hard on beneficial mites, but few other non-target species.

Copper products are strong fungicides and bactericides. Repeated use of copper over many years carries the risk of buildup of this micronutrient to toxic levels in the soil.

Hydrogen peroxide kills microbes on plant leaves and other surfaces. A related compound is hydrogen dioxide, the active ingredient of OxiDate, ZeroTol™, and StorOx®. While it is labeled for postharvest treatments, the manufacturer claims that it is legal to use StorOx® in field applications as a fungicide. There are varying reports as to the effectiveness of these products.

Surround® is a processed kaolin clay product that can be sprayed to coat plants with a white, repellant layer. It is very effective against some tree fruit pests such as plum curculio and pear psylla. It is somewhat effective against striped cucumber beetle but must be reapplied often.

Microbials

Helpful microbials include many products with live microorganisms that can be used to stimulate active pest resistance, or which may themselves be active against pathogens or arthropods. Entomopathogenic nematodes directly attack some insect pests (see the cucurbit section in chapter 7).

Compost teas need more investigation. Currently some studies show considerable potential with this farm-produced or purchased product, but others show negative results. The status of compost teas under the NOP is currently unclear (2003). For more on compost teas, see "Compost Teas for Plant Disease Control" and "Notes on Compost Teas," on the ATTRA Web site, http://attra.ncat.org, or have them mailed to you by calling (800) 346-9140.

Bt or *Bacillus thuringiensis* products contain insect toxins derived from various strains of Bt bacteria. Different strains can be highly effective against caterpillars and some fly and beetle larvae. They have a short residual period. Some are produced with genetically engineered bacteria and are not allowed in organic production.

Spinosad is the active ingredient of a new product called Entrust®. Like Bt, it is a microbially produced insect toxin. Spinosad has been in use for several years in a nonorganic formulation called SpinTor. It is very effective against caterpillars, providing an alternative to Bts, and also against Colorado potato beetles (adults and immatures), blueberry fruit flies, thrips, leaf miners, and, according to university trials, flea beetles (not currently on the label).

Botanicals

Botanicals are pesticides derived from plants.

Neem tree products are said to repel some insects and interfere with their proper development from larvae to adults. Neem oil also has fungicidal properties.

Rotenone is a moderately toxic contact and stomach poison. At this time, no products containing rotenone are on the OMRI list. Rotenone has short residual activity. It is effective against flea beetles and Colorado potato beetle larvae. It will kill some cucumber beetles, especially if dusted on when dew is on the plants. Rotenone is not very effective against caterpillar pests on crucifers.

Pyrethrum is effective against many flying insects and is hard on beneficials. It has short residual activity.

Garlic is used as a repellant. Most university studies show it has little effect on insect pests, but there is good evidence of anti-fungal properties.

Hot pepper products contain capsaicin as an active ingredient. Capsaicin is moderately effective as a flea beetle control and deer, woodchuck, and mouse repellant.

Plant oils including neem, sesame, mint, rosemary, and others show mild insecticidal and fungicidal properties.

Using Approved Sprays

Here are some tips on spraying:

- Use a good sprayer that allows for coverage inside the crop canopy.
- Often pesticide rates are given in pounds or weight (not fluid) ounces per acre. Weigh a cup or pint of the product you are using, and figure out a conversion from weight to volume. Do not assume that a "pint weighs a pound." Wettable powders often weigh around 8–10 oz. per pint.
- An organic spray product may not have the strength of a comparable conventional product. But if its use is supported by good cultural practices, it may still be effective enough.

Adjuvants

This category includes stickers, spreaders, penetrating compounds, and other additives that can make pesticides more effective. Stickers, like Nufilm 17, bind the spray ingredients to the plant surface. Some also protect from ultraviolet radiation, so applied pesticides last longer. Spreaders, typically surfactants, allow pesticide sprays to cover the leaf surface more completely. Penetrating compounds such as fish oils can carry some compounds into the leaf tissues (which can lead to phytotoxicity problems in some cases). These products, like any used in organic production, are subject to approval under the NOP.

Special Mammal and Bird Section

Mammals and birds deserve their own special discussion. They have big brains, at least compared to insects and fungi.

Deer are the ultimate herbivore in our region (I would never want to deal with elephants!). Deer pressure is highly variable from place to place, and different control measures may or may not be effective depending on the pressure.

In an ideal world, a few strands of baited electric fence would be enough to protect crops from deer. And this does work in some areas of low to moderate deer populations. Baiting is a key component of this type of fence, because deer must be enticed to touch the fence with their nose. Peanut butter in foil is a common choice, but commercial deer baits are available that last longer, are simple to refill, and easier to use. Fencing must be in place before deer find the crop and start to feed. These systems are relatively inexpensive and easy to move from one field to another to protect sensitive crops. Some growers have had success with plastic mesh fences, but in other cases deer have run right through them.

In areas of moderate deer pressure, the two- or three-strand electric fence with a baited "outrigger" single strand about 4 feet away will work reasonably well. Use a permissible deer repellant product like Miller's Hot Sauce on the inside perimeter to reinforce the fence. If possible, provide deer with alternative grazing forage like alfalfa or clover in an area away from your crops.

Sadly enough, though, in heavy-pressure areas the only sure defense is an 8-foot high, woven-wire fence. This costs around $6 per running foot (2003 prices) installed. So, for a typical 10-acre field with a 3500-foot perimeter, this is over $20,000, or $2000 per acre. As shocking as this number appears, in high deer pressure areas it is a wise investment. Do this before losing thousands of dollars worth of crops (you'll have to eventually do it anyway, or move)—bite the bullet, borrow the money, and invest in the fence. It should last twenty years. Amortized costs will be around $100 to $200 per acre per year, depending on interest rates. This is a good long-term investment.

In addition, encourage hunters on your land. Some states have nuisance permit programs to shoot marauding deer out of season. Check with your state environment office. Let them know if deer threaten your livelihood.

After deer, other warm-blooded critters seem downright easy. Woodchucks and rabbits are discouraged by a wide, well-mowed swath outside your fields. Spray your outer rows with hot pepper. A low "Electronet" (electrified plastic netting) fence around trouble areas is effective under heavier pressure. This also works for coons in the corn. Mice and voles are worst near the edges of fields next to unmowed meadow, and also can get into mulch and spread throughout the field. Keep an area unmulched around the outer edge of the field. Active dogs kept outdoors will do wonders against all mammal pests.

Birds are usually worst in sweet corn. Blackbirds and crows will peck out new seed and seedlings, going right down the row. Amazingly enough, anecdotal success is reported with placing small piles of old seed or feed corn on the edges of the fields. It is much cheaper to sacrifice this corn than the seed in your field! Birds can also be troublesome in strawberries and blueberries. Scare guns during peak season are probably the best repellants for these. Some growers string fishing line zigzagging two feet above the crop; another favors helikites (http://www.biconet.com/birds/helikite.html), which are a bit pricey. Scare-eye balloons that are placed in the crop that is just about to be harvested repel birds from that planting. Recorded distress calls also deter birds, but should be moved frequently. Keys to success with scare devices are using multiple modes (visual + auditory) and keeping 'em moving.

Birds help in the garden too. Encourage them in the great majority of crops for which they are beneficial. Shane Labrake of Ecofarm in Maryland puts fiberglass fence posts out in his crops. They serve as perches for foraging birds. In recent years, Colorado potato beetles have not been a problem for him, perhaps because of bird predation (personal communication).

Marketing to Reduce Pesticide Use

Many times, farmers spray pesticides on crops for purely cosmetic reasons. A little scab on an apple or potato makes no difference to its eating or nutritional quality; but it makes a huge difference in the price that a farmer can get for it on the wholesale market! This is because when a bin of slightly blemished produce is placed next to one of perfectly clean produce, the clean produce will sell faster (all other things being the same). But all other things are not necessarily the same.

Where the farmer is in direct contact with the consumer, she can explain and even demonstrate other aspects of quality besides appearance. Slices of odd-looking heirloom tomatoes for tasting quickly make believers out of many buyers. CSA shareholders gladly accept out-of-spec produce that wholesale buyers would turn away. And at farmers' markets, *slightly* blemished certified organic apples may command strong prices and sales, even in the midst of near-perfect looking conventional fruit.

The consumer is becoming aware that quality goes beyond simple appearance. It also includes nutrition, flavor, keeping quality, freshness, and wholesomeness. By reinforcing that fact with the customer, the farmer reduces the demand for purely cosmetic sprays.

If only we can do the same thing for the cosmetically imperfect lawn, the earth will heave a sigh of relief!

So, sometimes the use of pesticides can be reduced by good marketing. But the fact remains, we draw the line at any blemish that is repulsive or results in reduced shelf life.

Food Safety

Food safety is an important issue. Microscopic pests such as pathogenic *E. coli* or other food poisoning germs that can make our customers sick must be eliminated.

Food safety can be ensured by good hygiene and common sense. *E. coli* outbreaks have occurred when produce became contaminated by animal feces. Physically separate your produce handling, washing, and shipping areas so there is no chance of contact with manure. Wash water must be potable. Make sure your workers stay home when they are sick and that they have ready access to restroom and washing facilities as they work. This can mean portajohns with sanitizer dispensers in the field. Make sure your irrigation water is from a clean source.

NOP regulations on the management of manure and compost are extremely strict and, when followed, virtually eliminate these as a source of concern on an organic farm.

If possible, for produce that is washed, keep the temperature of the first wash about the same as the produce. If the first wash is very cold, tiny particles of dirt and bacteria can actually be drawn into the produce tissues. By waiting until a second wash to cool the produce, this problem is avoided.

four

Design Your Vegetable Farm to Reduce Pests

Farm design can involve everything from buildings and enterprises through physical layout and facilities to crop mix, crop rotation, and fertility practices. This chapter will focus on crop mix and rotation for a vegetable farm, and how such design choices relate to insect and disease pest pressure.

Diversity

The more crops you grow, the more potential pests you must deal with. The more individual plantings you have, the less time for pest scouting and management you can devote to each one. Do these add up to an untenable situation for the diversified vegetable grower?

Not necessarily, for this reason—the very diversity of the farm gives you more tools to foil pests. In-field diversity, spatial layout, and crop rotation are important tools that can greatly reduce pest pressure if used for that purpose. But planning must be done with care.

Crop Rotation–the Organic Farmer's Most Powerful Tool

Crop rotation is not only crucial for insect and disease management, but it is also vital for good weed control and soil nutrient management. Please refer to other manuals in this series for practical information on those other aspects of crop rotation. The rotation methods presented here are fully compatible with good weed and nutrient management.

Many organic farmers use a "catch-as-catch-can" approach to their rotations. Each year while planning their season, they consider a given crop, look for a field where it hasn't been grown for a few years, and write it in for that year. Invariably, some crops are shunted onto fields that are not ideal because of the haphazard nature of this approach.

This manual offers crop-rotation planning methods that group crops and place the rotation on a more regular patterned basis.

The initial planning is a bit complex, but when the overall rotation plan is set up, crops and fields fall into place each year. Allowance is made for the constant adjustment of crop mix that happens every season. Remember, the goal is to eliminate the Big Niche.

The first section discusses intensive vegetable rotations, in which each field grows a vegetable cash crop each year. The second covers extensive vegetable rotations that are far simpler and have many advantages.

These rotations are based on good scientific principles, logic, and

Vegetable-to-Vegetable Rotation Rules

Here are three basic rules for vegetable-crop rotation:

1. Allow an interval of at least three years between crops of the same family on the same field. This means at least a four-year rotation, and therefore, at least four different crop groupings. *Exception: This rule does not apply to the workhorse grass and legume families commonly used as cover crops. These should have at least one intervening crop between plantings of the same family.*
2. Don't plant succession plantings of the same family next to each other. At the least, break them up with strips of unrelated crops.
3. Adjacent crops in the rotation should be as widely unrelated as possible. The Solanaceae, Rosaceae, and cucurbits are fairly closely related, and should be separated by more than one year if possible. Grasses and the onion family are more distant, and good choices for intervening years; crucifers, umbellifers, composites, and beet/spinach family are intermediate choices. Some pests, like tarnished plant bug and sclerotinia fungus, have wide host ranges.

common sense, but have never actually been tested in field trials and shown to be better than the haphazard approach.

Farms in which each field is composed of many permanent beds have greater flexibility in their rotations and the opportunity to introduce high levels of diversity into their fields.

Rotation Crop Groups

As specified in the rotation rules listed above, a key goal of the rotation is to allow for a period of at least three years to intervene before a crop of a given family is planted again in a field. This is primarily for disease control. Many (but not all) important crop-disease organisms are transferable only within a given botanical crop family, and will die out in the field if crops from other families are grown. Weeds from the same botanical family must also be controlled for this to work.

However, a rotation in which whole fields are planted only to crops of a single family is not utilizing the ecosystem factor of diversity within the field. On most farms, it probably is best to group two or more crop families together in the rotation, rather than planting big blocks of only one type. This increases within-field diversity. Group together crops handled in similar ways, and grown in a similar season. This makes management easier.

Rotation Crop-Group Examples

Here is an example of a four-part rotation using crop groups and harvest times.

Group A. Short-season successional plantings—lettuce, spinach, herbs (umbellifers and mints), beets, beans, peas, crucifer greens.
Group B. Warm, long-season—cucurbits, corn, sunflowers, Solanaceae.
Group C. Fall-harvest plantings—crucifers, carrots.
Group D. Cool, long-season, and overwintered—alliums, potatoes, strawberries, garlic, parsnips.

These groupings are given as an example—you'll want to adapt them for your farm. Note that it gets tricky to obtain three intervening years

Botanical Families of Vegetable Crops

Including some common weeds (in brackets) and annual flower crops (in braces).

Broadleaf plants (dicots)	
Solanaceae	tomato, potato, eggplant, pepper, tobacco, husk cherry, tomatillo, [nightshade, horse nettle], {petunia}
Crucifers	brassicas (cabbage, broccoli, cauliflower, kale, collard, napa, Brussels sprouts, broccoli raab, mizuna, kohlrabi, etc.), radish, daikon, turnip, rutabaga, cress, [winter cress, yellow rocket, wild mustard]
Cucurbits	squash, melon, pumpkin, cucumber, gourd, bitter melon, calabaza
Umbellifers	carrot, parsnip, dill, cilantro/coriander, celery, celeriac, fennel, parsley, [wild carrot]
Composites	lettuce, chicory, radicchio, artichoke, burdock, endive, escarole, Jerusalem artichoke, [dandelion, galinsoga, ragweed], {sunflower, cosmos, marigold, zinnia}
Mints	basil, mints, oregano
Polygonaceae	buckwheat, rhubarb, sorrel, [lady's thumb, wild buckwheat]
Legumes	clovers, bean, pea, fava, cowpea, soybean, alfalfa, sweet clover, vetch
Goosefoots	beets, spinach, chard, [lamb's quarters]
Rosaeceae	strawberry, brambles
Miscellaneous	[purslane, chickweed, pigweed, bindweed]
Monocots	
Alliums	onion, leek, garlic, shallot, chives
Grasses	group 1—corn, millets, sorghum, Sudan grass group 2—ryegrass, small grains, turf grasses, [quack grass, foxtails, crabgrass, etc.]
Miscellaneous	asparagus, [yellow nutsedge]

between related crops if they are included in more than one group (for instance, crucifers in groups A and C). In this case, arrange your plantings so that when group C comes to a given field, the crucifers are in a different area than they were when group A was there. Because of the several crop families in each field, you can usually do that.

Setting up your rotation can be like solving a giant puzzle!

Equal Areas

In order to produce fairly consistent amounts of a given crop from year to year, each rotation group must be grown on about the same area. This is a problem on most farms, because for economic and marketing reasons, some groups might take up quite a bit more space than others. For instance, one farm might grow more sweet corn (group B) than anything else, while another might be heavy on lettuce (group A). Also, the market dictates that the areas of various groups change somewhat each year.

Allow some flexibility. To get an equal area of each group, change the crop groupings, or add cover crops as needed to "fill up" the area:

Warm-season compatible cover crops—Sudan grass, millets, soybeans, cowpeas.
Early-season—oats, peas, favas.
Late-season—oats, peas, rye, vetch.
Short-window—buckwheat, Japanese millet.

You can also split a large group into parts: For instance, divide B into B1—corn; and B2—cucurbits and tomatoes. This will add another season to the rotation.

You can combine groups, as long as you still have enough groups for a four-year rotation. For instance, add group B and D crops together to make a larger total area.

Example: Suppose that you have 6 acres of vegetable cropland. You want to grow approximately the following amounts of crops, to satisfy your markets and income needs:

Group A (successional crops)	1.5 acres
Group B (warm-season)	.5 acre
Group C (fall-harvest crops)	2 acres
Group D (cool, long-season)	0.75 acre

This leaves about 1.25 acres open to be planted in cover crops (CC) to complete the 6 acres. Let's add cover crops and rearrange them into four new groups of 1.5 acres each.

Group A	1.5 acres
Group B + D + 0.25 acre cover crop	1.5 acres
Group C1 (most of C)	1.5 acres
Group C2 + 1 acre cover crop	1.5 acres

How to Divide Up Fields to Fit Your Rotation

Most farmers have fields of various sizes. Does this mean that a given crop or crop group fluctuates in acreage as it rotates from field to field? No. If the crop-rotation groups occupy equal areas, it follows that the farm fields must be divided up equally.

Large-scale farmers often try to rotate their fields in groups, so that the total acreage of each group is about the same. This allows them to produce about the same mix of crops from year to year. Smaller-scale farmers can break up their fields and even group parts of them together to suit their rotation. For instance, here is an easy example. Suppose you had the following five fields that totaled 6 acres:

North 1—1 acre
North 2—1 acre
East—0.5 acre
West—1.5 acres
South—2 acres

Since there are four rotation crop groups in our example, let's divide these five fields up into four chunks of 1.5 acres each:

N1/S	North 1 + .5 acre of S	1.5 acres
N2/E	North 2 plus E	1.5 acres
W	West	1.5 acres
S	South less .5 acre	1.5 acres

Now, we simply assign the rotational crop groups to the 1.5-acre chunks of the fields. We are not bound to any particular order—just whatever fits together best. In this case, by placing group C crops (fall-harvested crops) following group D (cool, long-season crops), there is always time to harvest overwintered group D crops like strawberries and garlic before planting the late group C crops in July and August.

Once the rhythm of a good rotation is established, planting decisions flow smoothly.

In this 6-acre, four-year rotation example, as long as each crop family takes up less than 1.5 acres, plantings of related crops will have three intervening years of unrelated ones. However, if any family takes up more than 1.5 acres (¼ of the total acres), then the rotation is becoming unbalanced. The farmer needs to think hard about reducing its area or getting access to more land.

Keep the backbone of your rotation relatively simple. You can adjust things here and there within groups or fields, but stick with the basic, fairly simple plan.

Each year, you will want to adjust your crop mix to respond to markets and your management. In many cases, your rotation scheme will accommodate such changes. If you begin to see that your rotation is becoming unbalanced—i.e., too much of one or two particular crops—then you might need to find more land or rethink your crop mix.

Some seasons may not cooperate with your all-too-mortal plans! Heavy

Field	Year 1	Year 2	Year 3	Year 4
N1/S	B + D + 0.25 CC	C2 + 1.0 CC	C1	A
N2/E	A	B + D + 0.25 CC	C2 + 1.0 CC	C1
W	C1	A	B + D + 0.25 CC	C2 +1/0 CC
S	C2 + 1.0 CC	C1	A	B + D + 0.25 CC

rains at the wrong time or a quick market decision may raise havoc with your rotation and field plans. But don't worry—if the basic backbone of your plan remains intact, you can break the rules a bit.

If you have a field that is almost always wet in the spring, you may have to figure out a special rotation for it that does not include early spring crops (or install tile drainage). You can also get fancy and have two separate rotations to best utilize different soil types, or for crops within or outside of a deer fence, or for U-pick crops near the road, etc.

These well-planned rotations will perform much better over the long term than the catch-as-catch-can versions, characterized by questions like, "Well, where will we put the onions this year?"

Extensive Rotations

Most vegetable farmers of any type tend toward intensive production, meaning that vegetable cash crops are grown in every field every year. Some farms do quite a bit of double-cropping as well.

Intensive cropping produces the largest dollar value of cash crop from the least land. It arises from the farmer's perspective that land is limited, so he must cram as much production as possible into it. Sometimes this is true; other times farmers have more available land nearby than they realize.

A disadvantage of intensive systems is their tendency to build up pests. They also are hard on the soil and tend to require a steady diet of compost or other nutrient sources to maintain production.

Extensive rotations, on the other hand, give the soil more rest between vegetable crops. In extensive rotations, grains, forages, cover crops, and green manures are important components of the crop mix in addition to the vegetables. In many cases, it is possible to choose such crops so that they are not hosts for the same pests that affect your crops. They can provide an inhospitable environment for pests, and a good one for beneficials. In the mean time, they can improve soil health and fertility while they reduce weeds! Incorporating grains, forages, covers, and green manures as major parts of a vegetable-crop rotation makes the rotation less intensive compared to a vegetable system that uses nearly every field for vegetable crops. If one does not grow a diversity of vegetable crops, then incorporating such noncash crops into the rotation is a must.

As an example, take a simple three-year field-crop rotation:
Corn → Small Grain/Hay → Hay → (*repeat*).

It could be modified for vegetables to this four-year rotation:
Veg. Group 1 → Small Grain/Hay → Hay/Cover Crop →
Veg. Group 2/Cover Crop → (*repeat*).

An extensive vegetable production system that includes field crops and cover crops is probably best from the standpoint of soil fertility, soil health, and pest management, including weed control. As Anne and Eric Nordell (2000) have shown in northern Pennsylvania, extensive methods allow for low-cost, low-labor crop production. The Nordells grow vegetables and cover crops on their fields in alternate years. In spite of (actually, because of) time put into heavy use of cover crops, overall crop labor is less than on intensive farms—mostly because weed-control labor is so much less. (See *Cultivating Questions*, their collected articles from the *Small Farmer's Journal*.)

Compared to continuously growing vegetables, this rotation injects a soil-building sod crop into the sequence, which would also interrupt weed and disease cycles.

In these extensive vegetable-rotation examples, vegetable crops make up about 50 percent of the rotation. As long as "Veg. Group 1" and "Veg. Group 2" above contain mostly unrelated crops, it is easy to maintain a three-year succession of unrelated vegetable crops, except for legumes. Many vegetable-crop rotation problems disappear when the production system is made more extensive.

Spatial arrangement and field layout

Considering items from chapter 2 and the above, plus practical considerations, we can keep the following things in mind when laying out how each field will be planted:

- Within reason, the smaller the individual plots are for each crop, the better. Pests will have more difficulty finding the crop. Plot size must be compatible with tillage, spray, and harvest methods.

The plots can be repeated to fulfill the total acreage needed, but preferably not in one big mono-field.
- Higher plant species diversity in the field is usually better.
- Separate successive crop plantings with as much distance as practical within a field.
- Strip cropping can accomplish these goals.
- Untilled areas should be included within the field (these are sources of carabid beetles, etc.).
- Small but sufficient areas, rows, or beds can be planted to "beneficial insectary" crops like mints, umbellifers, alyssum, and other long-blooming small-flowered plants. Also, an occasional flower can be substituted within transplanted crops (for instance, plant a cosmos or marigold plant instead of each fortieth tomato).
- Windbreaks may be beneficial on hilltops, but are not usually helpful in the valleys; they may slow crop drying and raise humidity, thus increasing disease.
- Planting in the direction of prevailing winds can help reduce leaf wetness duration and humidity within the crop canopy.

Sod strips at Blue Heron Farm, Lodi, New York. PHOTO BY BRIAN CALDWELL.

In addition, consideration should be made for reducing or restricting traffic to avoid compaction and maintaining aisles for harvest, irrigation, and spraying.

Combining these suggests some sort of "diversified strip cropping" as an ideal arrangement. Examples of working farms with this type of system are Blue Heron Farm in Lodi, New York, and Peacework Organic Farm in Newark, New York. They use permanent unraised beds with sod wheel tracks in between. Jean-Paul Courtens at Roxbury Farm in Kinderhook, New York, plants strips of crops eight beds wide, with sod driveways between the strips for harvest and spraying. Such layouts allow for small crop plots, high diversity within the field, plenty of flexibility in arrangement of crops, and untilled areas.

By limiting machinery-wheel traffic to small, defined strips, these systems preserve soil health as well. This in turn helps pest management by promoting healthier plants.

Trade-offs with these systems include loss of crop area and increased mowing time and equipment. For most crops, it is best to have a mower that will not blow clippings to the side; though sometimes, such as when mulching around large plants, blowing clippings onto the crop area can be desirable. Precision tillage equipment such as a spader or rotovator is required. Similarly, using a backpack model for spraying is more precise, though it takes more time than tractor-mounted sprayers with bigger tanks.

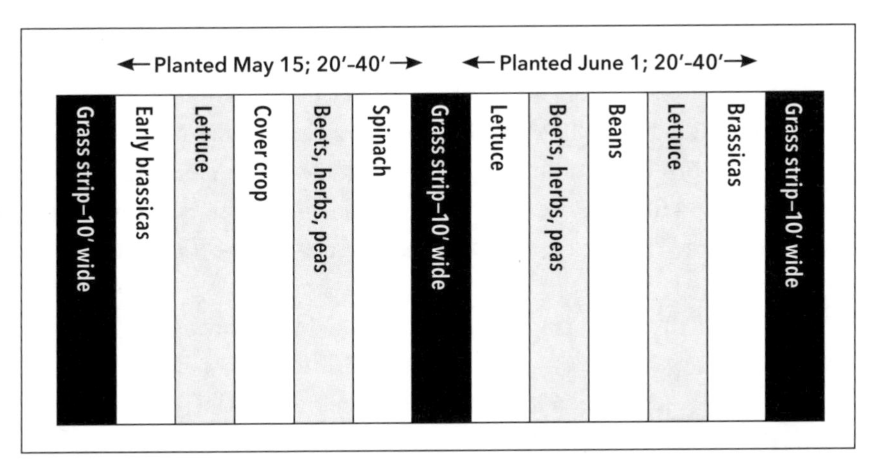

Strips of crops in a field with grass driveways every 20 to 40 feet.

This could be part of a field currently in rotation group A. The grass driveways allow for easy harvest or spraying access, plus they provide habitat for carabid beetles and other beneficials. The alternating-strip intercropping allows for more diversity within the field and helps confuse pests. Note that the crop strips can be varying widths to accommodate different amounts of crops in each planting. On the other hand, it would make management simpler overall if the distance between grass driveways was uniform over the whole farm.

Pitfalls of High Diversity

In some cases, increasing diversity with the wrong crop can actually increase pest pressure rather than help. As mentioned above, tarnished plant bugs (TPB) and some other pests have many hosts, so adding another crop that happens also to be a host can make the pest problem worse. For instance, if you grow hairy vetch, and allow it to flower and mature seed, you will be breeding thousands of TPB adults that will infest your other crops. This is because the TPBs will have time to complete an entire generation in the favored vetch host crop. If the vetch is tilled in at early flowering, which is the usual practice, the TPB nymphs in the crop will not have time to mature into adults and may be killed. Managing legumes in the rotation is tricky if low TPB pressure is needed.

Similarly, sod strips or mulched plots, while good for the soil and many beneficials, can increase problems with slugs or voles. If grass strips are nearby to crops that are subject to vole damage, keeping them mowed short is important to reduce vole habitat. Slugs may make it unfeasible to grow some crops under mulch. Jim Crawford of New Morning Farm in Hustontown, Pennsylvania, reports that when sod strips are within 100 feet of his raspberries or basil, he has serious Japanese beetle problems on the crop.

Once a basic overall farm design is established for a few years, problems will probably arise in spite of great planning. If the basic plan improves soil health, recycles nutrients, and fosters beneficial-insect habitat, it should be possible to change specific elements over time to correct them.

five

Unorthodox Approaches

Scientific Orthodoxy

Humility in the face of unexpected and inexplicable results is a virtue. Sometimes we agricultural "experts" need to remind ourselves of that, especially when we visit farmers trying seemingly outlandish methods.

Some approaches that were decried as scientific heresy twenty years ago are now hot ticket items for researchers, grants, and product development. An example would be "induced resistance." When I attended the Cornell University ag school in the 1970s, I was indoctrinated with the view that plants had no capacity to actively resist pests. In other words, once a plant was attacked by a pest organism, it could not change in any way to combat the pest. This was a dogma held by many plant pathologists and physiologists. To suggest that a plant could combat pests was to be accused of looking at the plant as if it were an animal, almost of being anthropocentric. It was even more radical to suggest that a plant could obtain generalized resistance by exposure to an attack.

Luckily, it was fairly easy to design experiments to show that plants exposed to a mild irritant (like copper ions, in an early case) would have lower damage under the same conditions as those that were not so exposed. (How these first experiments were ever funded, I don't know.) But after years of grudging acceptance, now several large and small companies

Disclaimers? Who Needs Disclaimers?

After all, modern science is now saying that 90 percent of the Universe is invisible; that 90 percent of microbes have been previously overlooked because they don't grow well on lab media; that 90 percent of an organism's genome is noise. . . . Perhaps there are still a few things we don't understand.

are touting products specifically formulated to induce resistance, such as Messenger®, Citrex® Lonlife, Actigard®, and others. These products are claimed to have properties verging on magic—increased yield and vigor in addition to pest resistance. While these products are not OMRI-approved, the approved microbial product PlantShield® may show this ability. Research on these products is now ongoing.

Farmers Out of the Box

It is not uncommon to find a farmer who chooses an unorthodox approach to pest management, and spreads the word about his or her success. This is the way that important advances are often made. In other cases, the approach doesn't catch on.

There is a tricky problem in generalizing from a particular farm. Sometimes a farmer is committed to a given method as the reason for pest-control success, but others who try that method fail. There could be several reasons for this: (1) the farmer wasn't as successful as he thought (may be selling to a more forgiving market); (2) other factors such as microclimate and general good management could be responsible for much of the success; (3) other farmers cannot master the subtle nuances of the methods without serious commitment, study, and trial.

To put it more bluntly, it could be that the method doesn't really work; something else is doing the job; or it's hard to implement the method properly. But in the hands of a skilled, persistent farmer, the method may be effective and confound our understanding.

Some novice-level methods will work for almost anybody. These form the basis of this manual. But following are techniques and philosophies that challenge scientific concepts and venture into rarified territory.

Biodynamics

Biodynamics is probably the most common organic farming approach that falls outside of conventional scientific thought. Based on the philosophy of Rudolf Steiner, biodynamics reflects a beautiful poetic mind-set.

Its farming practices center on ruminant farm animals, compost, special preparations, and celestial influences. The farm is viewed as a nearly self-contained organism. One unique biodynamic method of controlling pests is called "peppering." Peppering involves collecting and burning pests under the correct celestial influences, then dusting the ashes around the area to be protected.

Many outstanding farmers use biodynamic methods. Understanding this approach requires serious study, beyond the scope of this manual. Please see the "Resources" section for more on biodynamics.

Permaculture

The practice of permaculture goes beyond agriculture to include many aspects of an ecologically designed lifestyle. The classic work on the topic is *Permaculture One* by Bill Mollison and David Holmgren (1978).

Keith Johnson describes the basic philosophy: "As a system of design, Permaculture provides a new vocabulary for observation and action, attention and creation, that empowers people to co-design homes, neighborhoods, and communities full of truly abundant food, energy, habitat, water, income, and yields enough to share" (http://www.permaculture activist.net/). The intent of the permaculture approach is to design landscapes that "mimic the patterns and relationships found in nature" (Holmgren).

Cation Balancing

The late William Albrecht is considered to be a pioneer in organic farming research. His work at the University of Missouri led him to propose a theory based on the soil's cation exchange capacity (CEC). The CEC is the supply of readily available soil cations (including calcium, potassium, magnesium, sodium, and hydrogen) held by soil particles. Albrecht maintained that these cations needed to be available within certain proportions of the overall CEC for best soil health and crop growth. Proponents of this theory, including several outstanding northeast organic farmers, say that crops resist pests when the proper soil cation balance is achieved. A good discussion of the topic is on the Web at http://www.vabf.org/soilre1.php.

six

Identifying Pests

The Basics

In order to respond to a pest problem, it is crucial that the farmer identify the causal organism. Sometimes this is not easy!

Observe your crops. Look underneath the leaves, which is where the action is in the insect and disease world. Walk through your fields on a weekly basis, or more often. Keep your eyes peeled for anything that looks a bit odd in your crops. This could be off-color foliage, drooping leaves, frass (insect droppings), strange insects, leaf spots, bumps, holes—anything! When you see something a bit unusual, take the time to examine a few plants closely. This can save you big losses later on. Sometimes you will notice a problem while washing and packing a crop. Take the time to follow it up.

If you find damage, try to take samples of it to bring into the house and look at more closely. If there are insects associated with the damage, take them too. If you can't catch them, observe them closely to get a good description. Then head into the house to learn more about the problem.

I have some favorite references—*Diseases and Pests of Vegetable Crops in Canada* (Howard et al. 1994) is the best. The Penn State book, *Identifying Diseases of Vegetables* (MacNab et al. 1994) is also very good. These can be used as picture books, often narrowing down pest possibilities. However, be aware that identifying a pest by comparing it with a picture is far from foolproof! It makes sense to consult with professionals. In most areas, commercial growers can get free pest identification from their cooperative extension office. Use

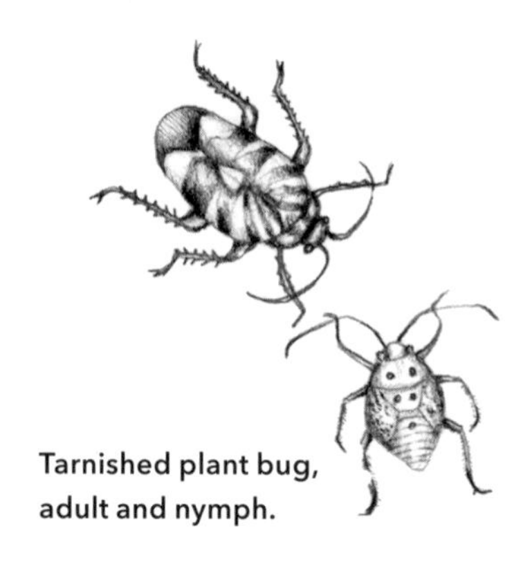

Tarnished plant bug, adult and nymph.

Carrot leaf damaged by leaf blight on tips of leaves.

that service! Nine times out of ten, the pest will be fairly well known, and easy to identify. The American Phytopathological Society has a series of compendia on vegetable diseases that are continually updated and complete (see http://www.shopapspress.org/).

Be aware that sometimes the insect you find in a hole is not the one that caused it! Your extension agent should be able to help you with this. The agent should have fact sheets on common pests as well, including valuable information on life histories.

It is particularly important not to blame damage on beneficial critters that may be feeding on the bad guys! Perhaps a good attitude in considering an unknown insect is "innocent until proven guilty." A great reference for beneficials is *Natural Enemies of Vegetable Insect Pests* by Hoffmann and Frodsham (1993). After reading through it, you will look at your crops with new eyes.

Once you have a positive identification on your pest, then you can learn about it and figure out what to do about it. The most common pests are mentioned in this manual under the crops they favor. Use the World Wide Web if you are so inclined, and you will be able to get all sorts of advice.

Be sure to learn about the life history of the pest. This will enable you to properly target and time any control measures. Perhaps the damage is already done or beyond control. This may happen when you encounter a new pest. Learn about it and be ready for it next year.

Special Situations

Nutrient Deficiencies

The lack of proper soil nutrients usually shows up as pastel discolorations of the leaves. Two good references on the Web that can help you pinpoint what your crops are lacking: http://www.hort.cornell.edu/commercial vegetables/online/product/nutrdefi.html and http://www.kmag.com/general/nds.htm (with pictures).

Luckily, if you have built up your soil fertility with compost or manure, and your soil pH is between 6.0 and 7.0, your chances of having nutrient deficiencies are fairly low.

Viruses

Virus diseases are tricky. Some viruses look like nutrient deficiencies, and some viruses may induce deficiencies. It is important to distinguish between deficiencies and diseases, because the remedies are so different. Knowing the virus will help you to find out its host range and vector—two very important aspects for management.

If a deficiency-like symptom is found on scattered plants with healthy plants around them, there is a good chance it is caused by a virus. But if many plants in certain parts of the field show symptoms with varying degrees of intensity, then a deficiency is more likely.

Mechanical Damage; Air Pollution

Sometimes plants may get damaged by equipment, row-cover abrasion, or some other similar agent. Or ozone levels may be high, damaging beans and other sensitive plants such as alfalfa, oats, onion, peanut, potato, radish, spinach, tomato, petunia, and grape. Ozone damage includes chlorosis, general bleaching of the upper leaf surfaces, and sometimes dying tissue.

Herbicide Drift

If crops along the edge of a field show symptoms of white spots or distorted growth, and drift could reach them from a sprayed field, the cause could be herbicide drift. Find out when and with what the neighboring crop was sprayed. If herbicide damage is likely, call your state environmental conservation department.

Sometimes Symptoms Will Stump Everybody

Years ago, some of our summer spinach leaves showed nasty brown areas from which the Cornell diagnostic lab could culture no pathogens. We never did find out what caused it. A favorite diagnosis when experts have no clue is "lightning damage"! Yes, it does sometimes happen, but it is rare. Look for a large circular area of dead plants. . . .

seven

Crop-by-Crop Pests and Practices

This is a general guide, encompassing many pest problems common to organic growers in the Northeast. However, you may encounter a specific problem that is not covered here. Get a good identification of the causal organism, then learn as much as you can about its life cycle, hosts, and favored conditions. See if you can find simple changes in your crop rotation or other cultural methods that will thwart it.

Not every pest mentioned here has a good organic-control approach. With your continued innovations and observations, however, someday they all will.

Asparagus

Insects

Asparagus beetle, spotted asparagus beetle. These beetles damage new spears and lay eggs on them. After the harvest season, their larvae can be found in the ferns.

Rather than the rescue treatment of spraying during the harvest season, it is preferable to reduce populations by targeting the larvae in the ferns. The approved pesticide Entrust® is labeled for this purpose. Spray the ferns about one to two weeks after harvest to kill most of the first generation. Some growers burn dead ferns after harvest to reduce insect and disease carryover.

During harvest allow some small spears to fern out. These ferns are more attractive to the beetles than the spears. Use the ferns as a trap crop and flame them on cool mornings when beetles move slowly. Or use row cover for early yields, some frost protection, and beetle protection.

> The adult beetles, which overwinter under plant refuse and debris along field borders, begin to move to asparagus as the

> spears first emerge in the spring. The beetles feed on the spears and lay eggs singly in vertical rows, usually near the tip of the spear. The eggs hatch in approximately 1 week, and the numerous fleshy, gray larvae move to foliage where they feed. The larval stage lasts 2–3 weeks, after which the larvae leave the plant, burrow into the soil, and pupate. Two or three generations are produced during the growing season in New York. (Cornell Guidelines)

Natural enemies of these beetles include several species of lady beetles. A parasitic wasp, *Tetrastichus asparagi,* lays its eggs in asparagus beetle eggs and is widespread in unsprayed asparagus plantings.

Diseases

Crown rot (*Fusarium*) is a major problem that tends to increase over the years. Consider establishing a new planting if it is getting severe. **Asparagus rust** (*Puccinia asparagi*) can also be serious. Keep plants healthy and vigorous, plant on raised beds in well-drained soil (this is critical for good asparagus production). Resistant varieties are available for both diseases.

Beans

Insects

The most common problems with early beans revolve around establishing a good crop stand. Early in the season, the **seedcorn maggot** may attack the seeds in the row as they sprout, greatly reducing the stand. This pest is the immature form of a fly that lays its eggs on top of the row. Row covers can be used to keep the flies out, and will also speed emergence by warming the soil. Besides presumably smelling the sprouting beans, these maggots are attracted to decaying organic matter, so avoid heavy applications of immature compost before early beans. Since beans fix nitrogen, they often grow well without added compost anyway. Check the beans a few days after planting to see if the seeds are firm and sprouting. If not, replant as soon as possible.

Various seed rots also affect beans during the early season. These can also be reduced by warming the soil with raised beds or row covers.

The **potato leafhopper** can be very destructive, especially if it arrives when the crop is young. Because potato leafhoppers migrate into the region, natural enemies that are present on the farm don't help control them. Adults and nymphs inject a toxin into the leaf when they feed causing "hopper burn." In beans, this causes rust-colored patches on the leaves, followed by yellowing and senescence. It can easily be confused with a disease; but leafhoppers can be seen on the underside of the leaves. Succession plantings to avoid infestations, row covers, and pyrethrum are control options.

In the warmer areas of the Northeast, the **Mexican bean beetle** (MBB) is a problem. Adults fly in and lay eggs on bean plants. The larvae then skeletonize the leaves. An effective approach is to make succession plantings, picking each one twice, then immediately tilling the plants under. Row cover (removed at flowering) may not be cost-effective against this pest, but can work. The parasitic wasp *Pediobius foveolatus* can provide effective season-long control as long as the initial MBB population is not too high. These wasps are available commercially. Please see http://www.caes.state.ct.us/FactSheetFiles/Entomology/fsen022f.htm. Neem and pyrethrum products are said to be effective.

Young bean plant being eaten by Mexican bean beetles.

Diseases

For **gray** (*Botrytis*) and **white** (*Sclerotinia*) **molds** maintain a good rotation, with grass species before the beans. Good air circulation and good soil drainage help. Avoid working in the crop or harvesting when the plants are moist. If possible, remove plants with white mold from the field, as the resting stage (sclerotia) of this fungus lives in the soil for years.

Beets, Swiss Chard

Insects

Spinach leaf miner can build up if hosts spinach, beets, chard, and lamb's quarters are present through much of the season. Long-distance rotation helps. Row covers can also protect against this pest. Remove any leaves with leaf-miner blotches during chard harvest and discard to dry out in the sun. The approved pesticide Entrust® is effective against leaf miners. Sticky tape may be effective (see chapter 3).

Diseases

Cercospora leaf spot is a common problem. At low levels, it is acceptable. Cornell Guidelines: Most prevalent in midseason with daytime temperatures of 75°F to 80°F and with frequent rainfall or long periods of 90 to 100 percent relative humidity. Copper products are effective against this disease, but of dubious merit unless pressure is known to be severe. Rotation, raised beds, and good culture are the best approaches.

Don't plant beets or chard before or after potatoes in a rotation, since all these crops are subject to the same scab root disease.

Carrots

Insects

Carrots have the same **aster leafhopper/aster yellows** situation as lettuce. See the lettuce section for control measures.

Diseases

Alternaria leaf blight, *Alternaria dauci*; **Cercospora leaf blight**, *Cercospora carotae*; and **bacterial leaf blight**, *Xanthomonas campestris* pv. *Carotae* are all seed-borne but can overwinter on plant debris on top of the ground and possibly in wild carrots. Diseased foliage may prevent carrots from being pulled from the ground. Good growing conditions, soil drainage, and rotation are the keys to keeping these at a low level. Some varieties have resistance. Raised beds are very helpful for carrots.

Celery, Celeriac

These plants need a lot of water for good production and quality!

Insects

Our friend the **tarnished plant bug** is back.

> Feeding on celery early in the season can cause severe heart injury. Late-season-feeding punctures on the stalks produce large, brown, wilted spots and a darkening of the tissue at the joint, causing an injury called "black joint." Natural enemies may help to control tarnished plant bug populations. They can be preserved by using insecticides that are less harmful to them. (Cornell Guidelines)

See the lettuce TPB discussion.

Caterpillars such as **cabbage loopers**, **fall armyworms**, and **beet armyworms** may be present. They are easily controlled with Bt products.

Diseases

Calcium deficiency can be confused with disease and can be associated with overfertilization with nitrogen (including from compost or manure) and subsequent reduction in calcium uptake. There is a good photo of this in the Canadian guide (Howard et al. 1994). Ruth Hazzard in Massachusetts reports seeing this on an organic farm where heavy fertilization of the bed seemed to have been the cause.

Sweet Corn

Insects

Seedcorn maggots. See the discussion under beans.

Flea beetles. A good rotation should take care of these. Remember, they are a different species than the flea beetles found on crucifers or the potato family. At low levels they are not a problem, but at higher levels they can transmit **Stewart's wilt** disease in significant amounts. Use resistant varieties.

European corn borer. ECB overwinters in the Northeast and will likely be present at higher or lower levels from May through September. Numbers tend to be higher in areas where lots of corn (both field corn and sweet corn) are grown. Use of row covers for early corn provides a barrier to ECB egg-laying and can be cost-effective, especially in conjunction with transplanting to get a good stand and extra-early corn for direct markets. Pheromone traps can be used to monitor moth levels, and field scouting can be used to determine the need for sprays at the pretassel stage. The best way to reduce them in your fields is to release *Trichogramma ostriniae* wasps when the ECB flight has started and the first corn is knee high. The wasps parasitize the egg masses and will reproduce through the season, knocking ECB levels down 50 to 90 percent, then die over the winter. Bt or spinosad (Entrust®) sprays can be used to control larvae. Contact your local cooperative extension office for instructions on how to scout your fields, release *Trichogramma*, and/or apply sprays, if you need especially clean corn. Learn how to quickly grade out infested ears, to provide your customers with a high-quality product.

Corn earworm. This caterpillar is a big obstacle for all growers. It overwinters in the South, and adult moths arrive in the Northeast from mid to late summer each year. Pheromone traps can be used to detect flight (captures of 2 moths per week indicate a damaging population!) or consult your local or state Extension office or a regional pest alert system such as PestWatch (http://www.pestwatch.psu.edu/) to determine when they are present. If moths are present, one can apply stylet, corn or soybean oil mixed with Bt to each silk with a hand-held applicator called a Zea-Later™, to achieve good control of this pest. Timing is important: oil should be applied when the block reaches 5-7 days past silk initiation. For more information, refer to the Biointensive sweet corn fact sheet from UMass, Johnny's Selected Seeds, or www.umassvegetable.org.

Diseases

If flea beetle numbers are low, Stewarts' wilt should not be a problem. **Common rust** is an occasional threat; it is best to grow tolerant varieties.

Cornell has ratings at http://www.nysaes.cornell.edu/recommends/26frameset.html.

Smut is also common in some sweet corn varieties. Again, from Cornell Guidelines: "Scout the field two or three times per season. Where feasible, cut out all of the smut balls before they break open, and destroy them by fire or burial. No thresholds are available. Young galls are considered culinary delicacies in some cultures. Consider marketing galls to upscale markets. Varieties vary in susceptibility."

Crucifers

This family includes broccoli, cauliflower, kale, collards, Asian brassicas, cabbage, kohlrabi, Brussels sprouts, radishes, rutabagas, daikon, turnips, arugula, etc. These crops can be considered as a single large group, since they are closely related and share the same pests.

Insects

Cabbage flea beetles are ubiquitous. They will stunt young seedlings or even kill them if numbers are high enough. They will also reduce yields and quality of larger plants if numerous. It is important to note that flea beetles on these crucifer crops are *not the same ones* that attack potatoes, eggplant, or corn.

Row covers are effective if put down immediately after planting. They need to be tight along the sides. Some leaf crops, such as arugula, are best grown this way.

Rotation can be effective against flea beetles, but is usually not because they can travel hundreds of yards, and are often present because of nearby cruciferous weeds. Such weeds include wild mustard, pennycress, yellow rocket, etc.

If crucifer crops are not grown before July and cruciferous weeds are well-controlled, then late plantings for fall harvest will have much lower flea beetle pressure. Organic mulch will also greatly reduce flea-beetle pressure.

Flea beetles can be reduced by dusting flour on the plants in the morning when they are wet with dew. Entrust® is effective against flea beetles, but not labeled for them. It is, however, labeled for several caterpillar pests on crucifers and will control flea beetles as well. At Cornell's Organic

Cropping Systems Project, we have found that a combination of Entrust® plus Surround® gives excellent control of flea beetles.

Trap cropping is a promising strategy for flea-beetle control. Hot mustards, arugula, and most oriental greens are much more attractive to flea beetles than, for instance, cabbage and broccoli. If seed is cheap, these trap crops could be planted before the main crop around field edges, then flamed or sprayed on a cool morning. A great many flea beetles could be destroyed that way.

It is possible that beneficial nematodes may help against this pest. See the cucurbit section for a discussion of these used against striped cucumber beetle.

Of the caterpillars, **imported cabbageworm** is the most common, but **diamondback moths** and **cabbage loopers** are sometimes pests. Bt products and Entrust® are effective. It is best to rotate use of these to prevent resistance buildup. In order to keep broccoli and cauliflower heads 100 percent clean, spray weekly from 1" diameter heads until harvest. Many beneficial wasps parasitize these pests, so having small-flowered plants in the field as a feed source for them is helpful.

Cabbage root maggots. Early plantings are the most susceptible to this pest. Spreading wood ashes around the bases of the plants is said to work. It will also help prevent clubroot disease. Row covers work well for early plantings. For perfect daikon radishes, turnips, and rutabagas, row cover or other controls should be used on late plantings as well. Sticky tape may be effective (see chapter 3).

Research has shown that in New York State, the cabbage-maggot-fly-generation peaks correspond closely with the flowering of common plant species—yellow rocket, daylily, early goldenrod, and New England aster. If the first generation is well-controlled and cruciferous weeds are not abundant, later generations of this pest are usually not a problem. Maggots may be present, but not in high enough numbers to reduce yields. Some growers skip early crucifer plantings entirely, or plant them only under row cover. Then later plantings are not bothered by this pest.

Aphids. Late in the season, aphids may build up in brassica crops. One grower suggests giving them an extra boost of boron to help them resist this. If this has been a problem on your farm, keep a close eye on aphid levels starting in mid-September, and try to stem the outbreak with a soap or plant oil pesticide before it gets out of hand. You will probably need more than one spray for a clean harvest.

Diseases

Major crucifer diseases include **Alternaria spot**, **black leg**, and **downy mildew**. A key aspect of managing these diseases is good rotation and air movement through the field. **Black rot**, a seed-borne bacterial disease caused by *Xanthomonas campestris* pv. *campestris* can be a problem. Separating later from early plantings is important, to keep diseases and pests from building up over the season. Clean seed and transplants help a lot. Copper sprays during wet periods will help. Be sure to use clean transplants. Optimize soil drainage with tiling and raised beds if necessary.

Fusarium wilt can be managed with rotation and resistant varieties. **Head rot**, which is worst on broccoli, is resisted by well-domed varieties. Copper sprays are somewhat effective. **Clubroot** is a serious problem, best prevented by keeping pH relatively high and rotating well. If a field becomes infested with the clubroot organism, it will be at least eight years before non-resistant crucifers can be grown there successfully.

Cucurbits

This family includes cucumbers, squashes, pumpkins, melons, and gourds. These crops are closely related and share pests.

Insects

Striped cucumber beetle (SCB) is the main problem. In addition to feeding voraciously on the crop, it carries bacterial wilt, which is devastating to many cucumber and muskmelon varieties, and some squashes and pumpkins. SCB overwinters as an adult in woods and hedgerows, and can quickly find cucurbit plantings in the spring. Row covers are effective. Trap-crop strategies also show promise for this pest. A listing of relative attractiveness of varieties to SCB is at: http://www.nysaes.cornell.edu/recommends/18frameset.html.

Likely trap-crop strategies involve planting highly attractive crops, such as dark green zucchini (an inexpensive OP bush variety that requires little space), before the main crop and either spraying or flaming it when SCB numbers on it become high. A perimeter planting of the trap crop can

Cucumber beetles eating cucumber seedling.

hold beetles out of the main crop, and is particularly effective when it can be sprayed.

Reels of yellow sticky tape (chapter 3) may be helpful also, over plants attractive to SCB. Surround®, a kaolin clay-based repellant/pesticide, is effective against this pest. In early growth stages, spray it twice a week to maintain coverage. It can be applied to transplants before setting out to save time and materials and ensure coverage before beetles arrive. Combining the use of trap crops, sticky tape, and Surround® may be the most effective solution short of row covers. Pyrethrum has not been effective against SCB.

Another experimental approach with striped cucumber beetles is to use beneficial ("entomopathogenic") nematodes. These are soil-dwelling nematodes that can attack those insects living part of their life cycle in the soil. SCB females lay their eggs by the stems of cucurbits, and the young larvae live in the soil feeding on the roots. A week or so after the beetles appear in significant numbers, commercial beneficial nematodes may be applied to the field to *reduce the following generation*. I have not read any studies of this practice, but it seems likely to work. *Heterorhabditis* or *Steinernema feltiae* are the best choices, as they actively move through the soil and can find SCB larvae. Nematodes must be applied under moist conditions—follow the directions.

Adult **squash bugs** are shy. They cause wilting of squash and pumpkin leaves, and particularly injure squashes of the Hubbard (*C. maxima*) group. Control measures for SCB, in particular row covers and PyGanic® sprays, help against squash bugs. In small plantings, boards can be used as a trap, checked early in the day, and bugs hiding under them destroyed. Adult squash bugs, and their offspring, are more numerous in mulched and no-till situations.

Ranking of Cucurbits by Cucumber Beetle Preference

Variety	Ranking[1]
Sunbar (SS, yellow)[2]	1
Slender Gold (SS, yellow)	2
Scallop (SS)	3
Seneca Prolific (SS, straightneck)	4
Goldbar (SS, straightneck)	5
Table Ace (WS, acorn)	6
Carnival (WS, acorn)	7
Yellow Crookneck (SS, crookneck)	8
Peter Pan (SS, scallop)	9
Baby Pam (P)	10
Munchkin (P)	11
Table King (WS, bush acorn)	12
Zenith (WS, butternut)	13
Tay Belle (WS, bush acorn)	14
Seneca Harvest Moon (P)	15
Butternut Supreme (WS, butternut)	16
Jack-Be-Little (P)	17
Jackpot (P)	18
Tom Fox (P)	19
Early Prolific Straightneck (SS, yellow)	20
Baby Bear (P)	21
Howden (P)	22
Spirit (P)	23
Wizard (P)	24
Early Butternut (WS, butternut)	25
Ghost Rider (P)	26
Big Autumn (P)	27
Waltham (WS, butternut)	28
Autumn Gold (P)	29
Jack-of-All-Trades (P)	30
Rocket (P)	31
Goldie Hybrid (SS, yellow)	32
Sundance (SS, yellow)	33
Sundance (SS, crookneck)	34
Frosty (P)	35
Spookie (P)	36
Multipik (SS, straightneck)	37
Connecticut Field (P)	38
Gold Rush (SS, zucchini)	39
Zucchini Select (SS, zucchini)	40
Ambassador (SS, zucchini)	41
Happy Jack (P)	42
Honey Delight (WS, buttercup)	43
Buttercup Burgess (WS, buttercup)	44
President (SS, zucchini)	45
Black Jack (SS, zucchini)	46
Big Max (P)	47
Cocozelle (SS)	48
Green Eclipse (SS, zucchini)	50
Seneca Zucchini (SS, zucchini)	51
Senator (SS, zucchini)	52
Baby Boo (P)	53
Super Select (SS, zucchini)	54
Ambercup (WS, buttercup)	55
Dark Green Zucchini (SS, zucchini)	56
Embassy Dark Green Zucchini (SS, zucchini)	57
Caserta (SS)	58
Classic (M)	59

Source: Cornell Guidelines for Vegetable Crop Production
1. The higher the number, the more preferred the variety by cucumber beetles. Rankings: 1–14 nonpreferred, >45 highly preferred.
2. C = cucumber; M = melon; P = pumpkin; SS = summer squash; WS = winter squash; W = watermelon

Squash vine borers are occasional pests of all pumpkins and squashes except the Butternut group (*C. moschata*). Oddly enough, they tend to be more severe in home gardens than in larger plantings.

Diseases

If SCB is well-controlled, **bacterial wilt** will be minimal to none. There are many **leaf spots** that affect cucurbits, but with good rotation and culture you should not see them often. ***Phytophthora* blight** is a major problem for pumpkins and squashes in some warmer areas of the Northeast. Be sure to grow them on well-drained soils, and rotate well. Peppers are another common host of this disease, strangely enough.

Powdery mildew (PM) will make an appearance each year on almost every farm. It comes in late July or early August, when plants are loaded with fruit. Unlike many fungal diseases, it does not need free moisture to get going. PM lowers yield and quality by reducing functioning leaf area. I believe that keeping soil quality very high, for a good, healthy root zone, will make plants more tolerant of PM. Choose spreading over bush varieties. Bush varieties have fewer overall leaves, so PM may affect their fruit more. If PM starts late and spreads slowly through the crop, its effect on yield and quality is minimal. Intercropping cucurbits with strips of other species can help with this. Try non-GMO PM-resistant varieties.

There are several approved pesticides that are effective against PM, but spray coverage must be very good, on tops and bottoms of the leaves. It is also difficult to get into the field when sprays need to be applied. Mineral oils, plant-derived oils, bicarbonate products, sulfur, and others will reduce this sensitive pathogen.

Virus diseases, including **cucumber mosaic virus** and others, can be a problem, especially in late plantings. The reservoir for these is perennial broadleaf weeds near the field. Aphids transmit the viruses to the crop. Either destroy the reservoir plants or keep cucurbits away from areas where the virus is prevalent.

Eggplant

Insects

See potatoes for ideas about **Colorado potato beetle** and **flea beetle** control. Early in the season, row cover works well against them for this high-value crop. The parasitoid wasp *Edovum puttleri* can control CPB on eggplant, but is ineffective in potatoes. Surround® (processed kaolin clay) is labeled for use against flea beetle in eggplant. Eggplants may lose substantial yield to **tarnished plant bug** damage to flower bud stems. Please see lettuce for a discussion of TPB. Avoid preceding or following eggplant with other hosts of ***Verticillium* wilt**—tomatoes, potatoes, and strawberries.

Garlic

Insects

There are no insect pests in our area that I know of; it is possible that **onion maggots** may build up in garlic if your rotation is heavily unbalanced toward the onion family. Be sure to use clean seed, well-drained soil, and a long rotation to avoid disease problems.

Lettuce, Chicory, Endive, Radicchio, Escarole, Italian Dandelion

Most of the pests mentioned below are worst on lettuce, but also will attack the other crops above.

Insects

Tarnished plant bug (TPB) is a major pest of upland lettuce. Adult feeding causes brown lesions on the midrib. Usually, it is worst from about mid-July to mid-August. TPB has hundreds of host species, wild and cultivated. It particularly likes to feed on flowers and tender flower stems. Common host crops include most legumes, buckwheat (when flowering), pigweed, brassicas, many rose family plants, etc. One common situation is that TPB may build up in legume hay crops, then invade nearby vegetable fields when the hay is mowed. About the only crops that are reliably not utilized by TPB are grass species.

In fact, this is a pest that may be best managed on a whole-farm basis. Theoretically, it should be possible to manage large trap-crop plantings such as hairy vetch as "sinks" for the TPB population. In other words, TPB adults are attracted to the trap crop and lay eggs there. After the trap crop has fulfilled this purpose, but before nymphs have reached a very mobile stage, the trap crop is mowed, tilled, flamed, or sprayed. Thus, TPB reproduction is reduced. Conversely, if a large host-crop planting

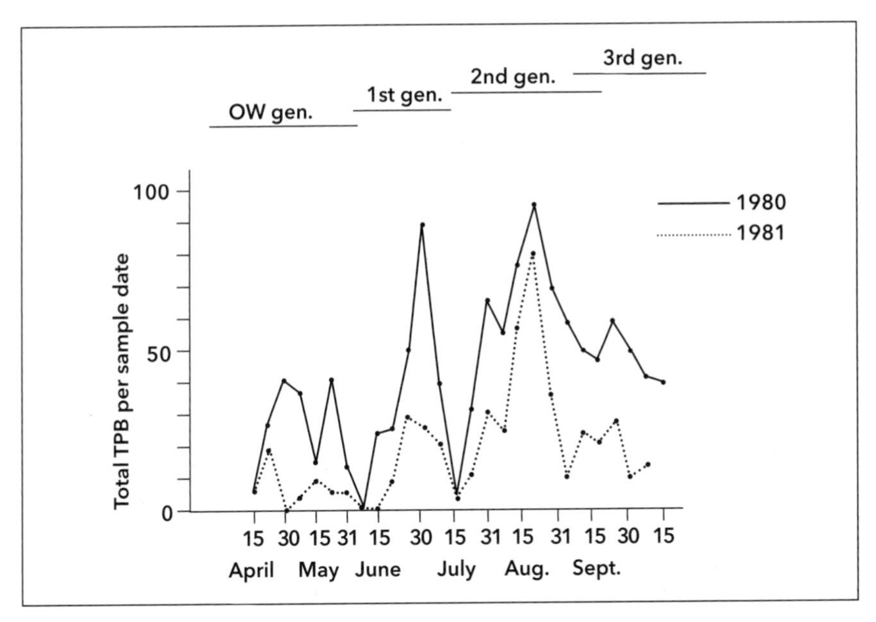

Catches of tarnished plant bug adults on weeds in and around peach orchard, Niagara Peninsula, 1980–81. (OW gen. = overwintered generation of adults. Chart from Stoner, *Alternatives to Insecticides for Managing Vegetable Insects*, NRAES, Ithaca, NY, 1999.) [diagram of TPB catches in Ontario] from: Roberts, W. P., and D. J. Pree (1983).

allows TPB to lay eggs that reach maturity, it becomes a source of more adults. The life cycle requires a minimum of three weeks during midsummer, and three or more generations per year are produced. TPB is much worse on some farms than others.

Natural enemies of TPB include big-eyed bug (*Geocoris punctipes*), and the wasps *Peristenus digoneutis, Leiophron uniformis, Anaphes ovijentatus, Peristenus pallipes.*

Row covers are not very useful here, since TPB attacks lettuce in the hottest part of the season. Lettuce quality will suffer under row covers during that time.

There is an excellent discussion on TPB in the book *Alternatives to Insecticides for Managing Vegetable Insects* (Stoner 1999). The discussion covers the many crops damaged by this pest, biological controls, and farmer experiences. In it, scientist Bill Day from New Jersey mentions TPB population peaks on June 20 and July 20. The chart shows midsummer peaks around July 1 and August 20 in Ontario, a cooler location.

Some growers avoid growing lettuce that matures during the late summer "TPB season." Romaine varieties tend to show the most damage; red leaf the least.

Aster leafhopper is not a serious pest in itself, but it carries **aster yellows** mycoplasma disease. A mycoplasma is a viruslike organism. Aster yellows incidence tends to build up in the second half of the season. It resides in overwintering weeds such as asters, dandelions, and Queen Anne's lace, then is transferred around by the leafhoppers. It does not overwinter in grass species, so mowing to reduce broadleafs will help. A diversity of crops in the field, as opposed to large plantings of only lettuce, may make it harder for the leafhoppers to spread the disease as easily.

Diseases

Gray mold (*Botrytis*) is best managed by using raised beds and not crowding the plants so that there is sufficient air circulation for them to dry out.

Bottom rot, downy mildew, and sclerotinia drop are all greatly reduced by long-term rotation, good weed control, and excellent soil drainage plus raised beds. Avoid irrigation during the last week before harvest. Here is more on bottom rot from the University of Connecticut IPM program Web site:

> Bottom Rot is caused by the fungus *Rhizoctonia solani*, which affects lettuce, escarole, endive, potato, pepper, eggplant, radish, cucumber, and many other fleshy plants. This worldwide disease was first identified on lettuce in 1900 in Massachusetts greenhouses. It is now a greenhouse and field disease and is favored by warm, wet conditions. Plants are usually affected when they are nearly mature.
>
> **Symptoms:** The first symptom seen from above is usually wilting of the outer leaves. Before this happens, the fungus enters the plant through lower leaves which are touching the soil. Slightly sunken spots, rust-colored to chocolate brown, appear on the leaf petioles and midribs. . . . The entire head may become a slimy brown mass that soon dries and becomes darker. . . . Avoid irrigation near harvest. Avoid rotating potatoes and other very susceptible crops with lettuce.

See the Rhizoctonia discussion in the potato section below.

Drop, *Sclerotinia sclerotiorum, S. minor.* The Cornell Guidelines have this to say about drop:

> **Time for concern:** The fungus is favored by warm, wet conditions and is primarily noticed at or near crop maturity.
>
> **Key characteristics:** The fungus attacks the outer leaves in contact with the soil, and wilting of these leaves is the initial symptom. Infection progresses inwardly until the entire plant wilts. Look for soft watery decay, snowy white mycelium, and seed-size black sclerotia (mustard seed for *S. minor* or pea seed for *S. sclerotiorum*).

The University of Connecticut IPM program recommends these preventative measures:

> Use long rotations away from lettuce, beans, celery, or carrots. The small grains are non-hosts for this fungus and are often included in rotations. Plant in well-drained soil and/or use raised beds. Steam greenhouse soil for one hour at 131°F or for 36 hours at 113°F. Space plants widely and avoid overhead irrigation to keep the soil surface dry. Flooding soil for 23 to 45 days destroys the resting structures of the fungus. Removing infected plants from small plantings is effective in preventing spread of the disease to other plants. Trim outer leaves after harvest, before packing, to avoid rot of the plants in storage.

Contans® WG consists of living spores of *Coniothyrium minitans,* a parasitic fungus that is said to destroy *Sclerotinia* in the soil.

Slugs can be a real problem when customers find them in the plants, though they do little damage to the crops in normal circumstances. There is a direct conflict between leaving high amounts of residue on the soil surface, which is good for the soil, and clean cultivation, which helps against slugs. Leaving mowed grass strips throughout the field can provide habitat for ground (carabid) beetles that eat slugs and their eggs. A new ferric phosphate slug killer is effective and nontoxic, but so far an organically approved formulation is not available.

Onions and Leeks

Insects

Onion thrips are important pests of onions, and leeks to a lesser degree. Their rasping damage provides sites for the entry of **botrytis, downy mildew** (*Peronospora destructor*), and **purple blotch** (*Alternaria*) fungi. It is thought that thrips are harbored on grass plants, including small grain cover crops, and may migrate onto onions. However, thrips have many natural enemies including lady beetles, minute pirate bugs, flower bugs, and big-eyed bugs. **Banded-wing thrips** (*Aeolothrips sp.*) are a genus of predatory thrips that is important in some situations. Nearby beneficial host crops such as clovers may provide enough predators to keep pest thrips populations from getting out of control. The above-mentioned diseases are reduced or avoided by good crop rotation, raised beds, and any practices that help the plants dry off quickly, such as orienting rows parallel to prevailing winds. Avoid planting onions near late broccoli, cabbage, and Brussels sprouts, as thrips may move from onions into brassicas. Spinosad (Entrust®), Pyrethrum (PyGanic® EC), and kaolin (Surround® WP) are labeled for thrips on onion.

Onion maggots, thank goodness, are not a problem on most organic vegetable farms. Good crop rotation generally keeps them in check. Be sure to properly compost cull onions to prevent them from allowing this pest to build up.

Parsnips

Be sure to get fresh, good-quality seed, as poor emergence and weed control are the big issues with parsnips. **Parsnip canker** can be troublesome, but many varieties have resistance. Otherwise, this is a fairly pest-free crop.

Peas

As with beans, poor emergence can be a big problem in cold soil. Fresh seed, raised beds, and row cover to warm the soil all help. Cultural controls are effective against pea problems. Rotate well, and don't work in the crop when it is moist. **Seedcorn maggots** can be a serious problem. Row covers will keep them out.

Peppers

Insects

Aphids transmit **cucumber mosaic virus** (CMV) and so can be a problem even in low numbers when CMV-infected hosts are near pepper plantings. CMV will greatly reduce yields and lower fruit quality. Keeping the periphery of your fields mowed to discourage perennial broadleaf weeds will help. Otherwise, **European corn borer** (ECB) is the main insect pest. It feeds on the leaves, then burrows into the fruit around the cap and feeds inside, making a mess. If you see "shothole" type damage on the leaves and the fruit is bigger than a quarter, a Bt or spinosad spray can kill the ECB larvae before they enter the fruit. Bts should be applied twice weekly and spinosad once weekly, until flights decline in late August. In southern New England, pepper maggot fly can be a problem. Flies are active in mid to late July and lay eggs under the cuticle, where maggots feed internally and exit the fruit to pupate. Pressure varies from farm to farm but if this is a problem, one strategy that can help reduce damage is a perimeter trap crop of two rows of hot cherry peppers around the main planting of bell peppers. Cherry peppers are preferred and will hold flies out of the main crop.

Diseases

Cucumber mosaic virus was mentioned above. Be sure that the greenhouse where your plants are grown is free of aphids, and has no overwintered plants, including weeds, that could carry CMV. Peppers are troubled by **bacterial spot** on some farms, particularly near intensive production areas. Resistant varieties are available. In warmer areas of the Northeast, **phytophthora** blight can be devastating. Paladin is a resistant bell pepper. Keep peppers away from squash and pumpkins in your rotation and ensure good soil drainage. Raised beds can help.

Tomato spotted wilt virus and **impatiens necrotic spot virus** may occur if transplants are grown with bedding plants. These tospoviruses are vectored by thrips.

Potatoes

Potatoes have a plethora of pests!

Insects

Potato leafhopper (PLH), though not so obvious as the Colorado potato beetle, may be a more problematic pest for organic growers. Adults are blown up from the South in early summer, some years in high numbers. Luckily though, in some years this pest is not a problem.

The adults lay eggs on potato plants, and the resulting nymphs do most of the damage. Both adults and nymphs inject a toxin into the leaves that causes sections of them to die rapidly. The plant "goes down" with "hopper burn," and tuber production abruptly stops. Many farmers don't even realize what the problem is.

Walk out into you potatoes and brush the leaves with your hand. If you see small white things darting around, you've got potato leaf hopper adults. Once you see them, look closely at several leaves. You will probably see small, elongated light green insects scurrying to hide from you—these are potato leafhopper nymphs.

PLH controls are not often used on organic farms, perhaps because growers don't realize how much yield loss this pest can cause. The adults do some damage, but the nymphs can really get out of control. Oils, soaps, and repellants such as neem oil seem likely to reduce them, but I am not familiar with results of trials with these materials. Surround® has been tested, and, unfortunately, does not appear to be very effective against PLH. Intercropping potatoes with strips of other crops should help, compared to large monocrop blocks. Avoid intercropping with beans, which are also very attractive to PLH. Later-maturing varieties including Katahdin, Elba, and Green Mountain, Kennebec, and Blossom have some resistance.

PyGanic® EC, an approved pyrethrum product, is reported to be effective against PLH. Two sprays of PyGanic® a week apart, as soon as you find significant numbers of adults in your planting (the IPM threshold for this pest is very low), may halt their damage and prevent a buildup of nymphs. Alternatively, you could tolerate the adult damage, then try spraying with a more benign material such as a plant oil product (soap; neem oil; Organocide™, a sesame/fish oil product; etc.) if the nymph population got too high. Hopefully, within a few years we will have a clearer, proven strategy against this pest.

Colorado potato beetle (CPB) is the widely recognized potato insect pest. Long-distance rotation of over a half mile is an effective deterrent, as is mulching with straw or other organic materials. A surprising number of growers still get kids to pick "potato bugs" into cans of soapy water. If your certifier allows it, whiz those beetles up in a blender with non-soapy water, strain, and spray them back on the plants. Any insect diseases they may have had will be spread throughout your planting, possibly to infect any remaining CPB. Some Bt products are effective against small CPB larvae, but currently none of these are on the OMRI list. Entrust® is labeled and effective against adults and larvae, however, if things should get out of hand.

The **potato flea beetle** is not the same insect as the one that attacks crucifers, and both are different from the corn flea beetle. However, the potato flea beetle will also go for eggplants and sometimes tomatoes. An organic mulch helps again. Please see the discussion under crucifers. Usually, potato flea beetles are not a serious problem, except on eggplant.

Wireworms are a problem when potatoes are planted after sod, so avoid that situation.

Diseases

Potatoes are liable to many diseases, some of which are physiological disorders like **hollow heart**, which happens simply because the tuber is growing too fast. Plant susceptible varieties like Chieftain a bit closer together to slow them down and get higher yields of reasonably sized spuds. Be sure your soil fertility is well balanced.

Common scab is a common problem. Scab and rhizoctonia are probably the main reasons that people used to peel potatoes. Both are skin diseases that look ugly. Large-scale, nondiversified growers can lower the pH of their potato ground below 5.2 for good scab control. However, this is not a good option for most diversified vegetable growers. A soybean green manure the year before potatoes is reported to control scab (Anon. 1966, 119). Use disease-free seed potatoes, preferably of varieties with good scab resistance.

Rhizoctonia. A green manure of Japanese millet, oats, or corn, plowed in 3 to 7 weeks in advance of the crop, is said to reduce this surface-blemish problem (Anon. 1966, 67). Sorghum Sudan grass and crucifer cover crops in advance of potatoes have also been suggested as good "cleaning crops" for this disease. "Rhizoc" shows up as raised black spots on the potato skin. These spots are sclerotia that allow the fungus to survive in the soil. Planting seed-pieces with sclerotia may lead to crown rot of the emerging potato plant. Unfortunately, the organism, *Rhizoctonia solani*, is a ubiquitous soil organism that can live off of decaying organic matter as well as live tissue. So, having a lot of new organic matter in the soil tends to make this pathogen worse. Recently, microbial products have come on the market that are antagonistic to rhizoc. PlantShield (*Trichoderma harzianum* strain T-22) has been effective in some, but not all, applications.

Late blight, caused by *Phytophthora infestans*, can be a disaster. The pathogen is restricted to potato and tomato. The spores may blow in from other areas many miles away, so rotation is of little value. Once the disease starts, it spreads extremely quickly. If you get it before September, the most probable source is infected seed potatoes. It may also survive on volunteer plants that were infected the previous year.

Luckily, late blight does not infect many farms in most years. It is enhanced by overcast and wet weather during the latter part of the season. Your local cooperative extension office should be aware of nearby outbreaks of late blight on tomatoes and potatoes. If this occurs, you need to decide whether you will apply a protective spray. Copper is effective with good coverage on a five-day schedule; or perhaps try a microbial product to cover the leaf surfaces. In most years, these sprays will be unnecessary (though the copper will also reduce early blight). However, late blight can cause 100 percent loss of unprotected crops. If late blight ravages your tomatoes or potatoes, the neighborly thing to do is to destroy that crop as soon as possible, to prevent spores from your farm from ruining crops on other farms miles away. Unchecked, late blight can produce billions of spores even from a small planting.

I believe that growers could get together and run a late blight self-insurance program, where they would contribute a little each year to a fund. Since only a few growers have major losses even in a wet season, they could be compensated for their crop loss out of the fund. In this way, growers could avoid protective spraying against late blight without risking their income from the crop.

For other diseases, use only clean seed and maintain good fertility and a good crop rotation. They should show up only infrequently.

Rhubarb

Insects

The main pest encountered with this crop is the **rhubarb curculio**, which makes holes in the stems. Curly dock is considered its favorite host, so eliminating it near your rhubarb should help. Some growers use row cover on rhubarb for an extra early harvest, which likely also protects early stems from this pest. Plant rhubarb on well-drained soil to escape disease problems. Small, well-fertilized plantings of this crop can be profitable.

Spinach

Insects

Spinach has few insect pests. **Leaf miners** (see beets) may be a problem, especially where many succession plantings of beets, chard, and spinach are made, or where lamb's quarters are heavy. Entrust® is labeled for it. There is a spinach flea beetle as well, but it is not very common. Recently, I've heard two growers complain that striped cucumber beetles are munching on their spinach! Good overall control of SCB should keep them from overflowing onto other crops (one hopes).

Diseases

Spinach diseases are much more of an issue. Spinach is a prima donna, requiring a high pH and more available nitrogen than it can use. It is subject to **downy mildew** (*Peronospora effuse,* aka blue mold) in early plantings. Some varieties are resistant. Note that this is not the same organism that causes downy mildew in lettuce or onions. It may well, however, be hosted by the closely related plants beets, chard, and lamb's quarters.

When temperatures are above 80°F, **cucumber mosaic virus**, known as spinach blight on this crop, can ruin a crop in a few days. Again, choose tolerant varieties. Be sure to rotate your spinach well.

Spinach is also subject to disease organisms that cause poor germination such as **fusarium** or **pythium** pathogens. Some growers report that these can build up in the soil until getting a good stand of spinach is nearly impossible. Those who have used treated seed in the past say that it helps, but conventionally treated seed is no longer allowed in organic production. PlantShield® or RootShield® applied to the seed may be helpful under these conditions. Otherwise, be sure that only very mature compost is used ahead of spinach. Another idea is to try a "cleaning crop" of a dense crucifer or marigold cover crop before spinach. These crops release natural soil fumigants when tilled in. Some growers grow multiplant transplants in potting soil, then transplant the crop.

Tomatoes

The single most important thing you can do to increase the marketable yield of tomatoes is—get them off the ground. Use whatever form of support you like, Florida weave trellis, caging, or staking, but do it.

Insects

Only a few insects bother tomatoes in the Northeast. **Tomato hornworm**, where common, can do a lot of damage in a short time. Find and kill the large caterpillars as soon as you see stripped leaves. It is quite susceptible to Bt, so use that if you have a severe infestation. More common and mysterious tomato pests are **stinkbugs**. They cause cloudy whitish patches under the skin of the fruit, which reveal an unripe, hard area when cut into. This damage definitely reduces eating quality. Stinkbugs are worst in small tomato plantings on the edges of fields. More research is needed on repellants or controls for these pests. Here is what the North Carolina State University Web site has to say:

> The green and brown stink bug can be important direct pests of tomato, but they are sporadic in occurrence. Stink bugs are most common in smaller fields (i.e., 5 acres or less) that are surrounded by weedy borders, or fields that are adjacent to soybeans. In fact, chemical control of stink bugs is oftentimes not necessary in fields that do not fit the above description. Unfortunately, there is not a good sampling method to assess population densities before damage is inflicted, and preventive strategies are used.

Diseases

Early blight is a yearly visitor on most farms. Indeterminate varieties that are well-fed, mulched with plastic, and staked (cages do the same thing)

The Benefits of Staking Tomatoes

Steve Reiners, Professor of Vegetable Science,
New York State Agriculture Experiment Station, Geneva

Why bother to stake your tomatoes? Researchers at Rutgers University have proven what many growers already suspected. Growing tomatoes upright on a trellis will lead to significantly greater yield of marketable fruit, less foliar disease, and fewer postharvest losses of fruit. In a two-year study, "Celebrity" tomatoes were planted on black plastic mulch. Half of the plants were allowed to sprawl on the ground and the other half were pruned, staked, and tied using the Florida Weave method. The results indicate that there was a 5-ton/acre advantage in marketable yield with the staked tomatoes as well as a significantly greater percent of marketable fruit. Foliar disease ratings were lower with staked tomatoes. In one year, 34 percent of the fruit from the ground cultured plants were lost six days after harvest to decay while only 10 percent were lost in the staked plots. The staked fruits only showed a 5 percent loss due to anthracnose while the ground cultured fruit showed a 25 percent loss. Why the difference? Staked plants have improved air circulation, which lowers humidity and decreases drying times. They also receive better fungicide coverage. Staking reduces soil contact with fruit and lessens potential splashing of disease organisms from the soil. Finally, staking results in less fruit cracking and weather checking, reducing potential infection sites. One word of caution, however. The increased handling of plants increases the likelihood of spreading bacterial diseases, especially in wet springs.

The study clearly demonstrated the benefits of stake culture for tomato production. Improvements in yield, foliar disease control and postharvest losses seem to more than make up for the increased cost of this production system.

are often able to tolerate this fungus reasonably well, producing good crops without sprays, until cool weather in mid-September finishes the plants. They continually make new leaves to replace those that go down to the disease, which may get off to a slower start if both rows and aisles are mulched. This keeps spores from splashing from the soil up to the foliage. The aisles between plastic need to be mulched with straw or hay for best effect. This is great for the soil, and prevents weeds.

If a spray is needed, copper products are effective. They must be applied on a seven-day cycle during periods of wet weather, or with an interval of up to fifteen days in dry weather. Predictive models such as Tomcast can be used to time copper sprays. They require close weather monitoring. Check with your cooperative extension educator for detailed information.

Work is being done with compost teas on this disease, but so far there are no clear guidelines for effective control that I'm aware of. Also, the status of compost teas under the NOP is unclear. Abby Seaman of Cornell has found that drenching the plants at transplanting time with PlantShield has reduced early blight in the field.

Septoria leaf spot is common in some areas. Its management practices are the same as those for early blight.

Late blight can affect tomatoes. See the discussion under potatoes. So far, there are no commercial late-blight-resistant tomato varieties, though Dr. Randy Gardener at North Carolina State has some in the pipeline.

Anthracnose is a late season problem. Mulching and staking your plants will greatly reduce it.

Bacterial diseases, especially **bacterial canker**, are endemic on some farms, usually in intensive production areas. Being extra scrupulous about greenhouse hygiene is important in reducing this problem. Copper sprays will help.

Other diseases are generally not important if good cultural practices and rotation are followed.

Postscript: The Future of Organic Pest Management

Organic farmers should be proud that they have furthered the practice of environmentally sound pest management, in an era future generations will look back on as the Toxic Time. Consumers should be proud that they have supported this effort. In the future, conventional farm pest-control practices will look more and more like organic ones—a healthy development. Researchers at many institutions around the world are looking for methods and materials that are much less environmentally destructive than the old standbys.

Countering these holistic efforts will be the narrow-minded but extremely well-funded push for genetic-engineering solutions to pest problems. This reductionist approach will lead to increasing, unanticipated negative side effects for the consumer and the environment. Let's hope that negative public opinion continues to put the brakes on this runaway technology. We do not have to accept this Brave New GMO World.

The challenge we face is to try to live in ways that allow a healthy future for our children. Making organic farming a great success is one small, demanding, exciting part of a giant effort.

Godspeed to all of you pursuing this goal.

Appendix

Several new pesticide products have recently become available that are approved for organic production and do not require an EPA label or New York State registration for legal use. The active ingredients of these products are on a special EPA list of exempt materials under the Code of Federal Regulations, Title 40, Sec. 152.25: "Exemptions for pesticides of a character not requiring FIFRA regulation."

Several ingredients potentially useful for organic pest management are on this list:

Castor oil
Cedar oil
Cinnamon and cinnamon oil
Citric acid
Citronella and citronella oil
Cloves and clove oil, eugenol
Corn gluten meal
Corn oil
Cottonseed oil
Dried blood
Garlic and garlic oil
Geranium oil, geraniol
Lauryl sulfate, sodium lauryl sulfate
Lemongrass oil
Linseed oil
Malic acid
Mint and mint oil
Peppermint and peppermint oil
Putrescent whole egg solids
Rosemary and rosemary oil
Sesame and sesame oil
Sodium chloride (common salt)
Soybean oil
Thyme and thyme oil

As long as a product contains only these active ingredients and is formulated with EPA list four inerts, it is exempt from EPA registration and also does not need to be registered for use in New York.

References

Books and Articles

Agrios, George N. 1988. *Plant pathology.* San Diego, CA: Academic Press.

Altieri, M. A., and C. I. Nicholls. 1998. Biodiversity, ecosystem function, and insect pest management in agricultural systems. In W. A. Collins and C. O. Qualset, *Biodiversity in agroecosystems.* Washington, DC: CRC Press.

Andow, D. A., and K. Hidaka. 1989. Experimental natural history of sustainable agriculture: Syndromes of production. *Agric. Ecosystems Environ.* 27:447–62.

Anon. *The organic way to plant protection.* 1966. Emmaus, PA: Rodale Press.

Boucher, J. 2003. Why perimeter trap cropping works. In *New York State Vegetable Conference Proceedings,* 2003. Kirkwood, NY: New York State Vegetable Growers Association.

Caldwell, Brian, Emily Rosen, Eric Sideman, Anthony Shelton, and Christine Smart. 2005. *Resource guide for organic insect and disease management.* Geneva, NY: New York State Agricultural Experiment Station at Cornell University.

Cheesman, Evelyn. 1952. *Insects: Their secret world.* New York: William Sloan Associates.

Coleman, Eliot. 1995. *The new organic grower.* 2nd ed. White River Junction, VT: Chelsea Green Publishing Co.

Cunningham, Sally. 1998. *Great garden companions.* Emmaus, PA: Rodale Press.

Gilman, Steve. 2011. *Organic soil-fertility and weed management.* NOFA Organic Principles and Practices Handbook Series. White River Junction, VT: Chelsea Green Publishing Co.

Henderson, Elizabeth, and Karl North. 2011. *Whole-farm planning: Ecological imperatives, personal values, and economics.* NOFA Organic Principles and Practices Handbook Series. White River Junction, VT: Chelsea Green Publishing Co.

Hoffmann, M. P., and A. C. Frodsham. 1993. *Natural enemies of vegetable insect pests.* Ithaca, NY: Cornell Cooperative Extension.

Kroeck, Seth. 2011. *Crop rotation and cover cropping: Soil resiliency and health on the organic farm.* NOFA Organic Principles and Practices Handbook Series. White River Junction, VT: Chelsea Green Publishing Co.

Lewis, W. J., J. C. van Lenteren, S. C. Phatak, and J. H. Tumilson, III. 1997. A total system approach to sustainable pest management. *Proceedings of the National Academy of Science* 94: 12243–12248.

MacNab, A., A. F. Sherf, and J. K. Springer. 1994. *Identifying diseases of vegetables.* AGRS-21. University Park, PA: Pennsylvania State University.

Marino, P. C., and D. A. Landis. 1996. Effect of landscape structure in parasitoid diversity and parasitism in agroecosystems. *Ecological Applications* 6 (1): 276–84.

Mishanec, John. 2003. Successful trap cropping for Colorado potato beetles. In *New York State Vegetable Conference Proceedings*, 2003. Kirkwood, NY: New York State Vegetable Growers Association.

Mollison, Bill, and David Holmgren. 1978. *Permaculture one.* Tyalgum, New South Wales, Australia: Tagari Publications.

Nordell, Eric and Anne. 2000, 2002 (2 vols.). *Cultivating questions*, collected articles from the *Small Farmer's Journal.* Available from the authors at 570-634-3197.

Phelan, P. L., K. H. Norris, and J. F. Mason. 1996. Soil-management history and host preference by *Ostrinia nubilalis*: Evidence for plant mineral balance mediating insect-plant interactions. *Environmental Entomology* 25 (6): 1329–36

Phelan, P. L., J. F. Mason, and B. R. Skinner. 1995. Soil fertility management and host preference by European Corn Borer, *Ostrinia nubilalis* (Hubner), on *Zea mays* L.: A comparison of organic and conventional farming. *Agriculture, Ecosystems and Environment* 56: 1–8.

Pickett, C. H., and R. L. Bugg, eds. 1998. *Enhancing biological control.* Berkeley, CA: University of California Press.

Roberts, W. P., and D. J. Pree. 1983. Tarnished plant bug. Fact sheet 83-027 Ontario Ministry of Agriculture and Food.

Root, R. 1973. Organization of a plant-arthropod association in simple and diverse habitats: The fauna of collards (*Brassica oleracea*). *Ecol. Monogr.* 43:95–124.

Ruutilla, E. 1999. Living straw mulch for suppression of Colorado potato beetle. In Kimberly Stoner, ed. *Alternatives to Insecticides for Managing Vegetable Insects.* Ithaca, NY: NRAES.

Sarrantonio, Marianne. 1994. *The northeast cover crop handbook.* Kutztown, PA: Rodale Institute.

Schonbeck, Mark. *Soil cation nutrient balancing in sustainable agriculture: Missing link or red herring?* Lexington, VA: Virginia Association of Biological Farmers.

Scriber, J. M. 1984. "Host-plant suitability." In W. J. Bell and R. T. Carde, eds. *Chemical ecology of insects.* New York: Chapman and Hall.

Stoner, Kimberly, ed. 1999. *Alternatives to insecticides for managing vegetable insects.* Ithaca, NY: NRAES.

The Sustainable Agriculture Network. 2007. *Managing cover crops profitably.* 3rd ed. Washington, DC.

Westcott, Cynthia. 1973. *The gardener's bug book.* 4th ed. Garden City, NY: Doubleday.

Web Sites

American Phytopathological Society. APS has a series of compendia on vegetable diseases that are up-to-date and complete; see: www.shop apspress.org/.

ATTRA. "Compost Teas for Plant Disease Control and Notes on Compost Teas," http://attra.ncat.org or have them mailed to you: 800 346-9140.

Cornell Cooperative Extension. 2003. Integrated crop and pest management guidelines for commercial vegetable production. www.nysaes .cornell.edu/recommends/.

Howard, Ronald J., Allan Garland, and W. Lloyd Seaman, eds. *Diseases and pests of vegetable crops in Canada.* Canadian Phytopathological Society and the Entomology Society of Canada, www.esc-sec.org/disease.htm.

North Carolina State IPM fact sheets, www.ces.ncsu.edu/depts/ent/notes/ Vegetables/vegetable_contents.htm.

OMRI www.omri.org.

University of Connecticut, IPM Program, www.hort.uconn.edu/ipm.

www.biconet.com/birds/helikite.html.

www.neon.cornell.edu/ or www.nofaic.org.

Resources

The books listed in this section represent just a few of those available. In my opinion they contain very reliable and useful information. Not all are aimed specifically at organic farmers, but all have a great deal of relevant information for them.

Vegetables

Grubinger, Vernon. 1999. *Sustainable vegetable production from start-up to market.* Ithaca, NY: NRAES.
Outstanding general resource. IPM/organic orientation.

Coleman, Eliot. 1998. *The new organic grower.* 2nd ed. White River Junction, VT: Chelsea Green Publishing Co.
Very good for smaller-scale producers.

Division of Agriculture and Natural Resources, UCSF. 1998. *Specialty and minor crops handbook.* 2nd ed. Publication 3346. Oakland CA: Communication Services—Publications, University of California San Francisco.
Many organic producers grow these crops.

Stoner, Kimberly, ed. 1999. *Alternatives to insecticides for managing vegetable insects.* Ithaca, NY: NRAES.
Excellent resource.

Cover Crops

Sustainable Agriculture Network. 1998. *Managing cover crops profitably.* 2nd ed. Beltsville, MD: National Agriculture Library.
Comprehensive.

Sarrantonio, Marianne. 1994. *Northeast cover crops handbook.* Emmaus, PA: Rodale Institute.
Very farmer-useable.

Weed Control

Bowman, Greg, ed. 1997. *Steel in the field.* Beltsville, MD: National Agriculture Library.
Essential for those using mechanical weed control. List of manufacturer contacts alone is extremely helpful.

Soil Management

Magdoff, Fred. 1992. *Building soils for better crops.* Lincoln, NE: University of Nebraska Press
Good discussion of organic matter management.

Field Crops

Michigan State University. 1998. *Michigan field crop ecology.* Ext. Bulletin E-2646. Lansing, MI.
Well-illustrated ICM approach.

Biodynamics

Biodynamic Farming and Gardening Association, Inc.
Building 1002B, Thoreau Center, The Presidio
P.O. Box 29135
San Francisco, CA 94129-0135
415-561-7797
415-561-7796 fax
biodynamic@aol.com
http://www.biodynamics.com/
Publisher of Biodynamics *bimonthly journal, Kimberton agricultural calendar, and extensive selection of books.*

Josephine Porter Institute for Applied Biodynamics, Inc.
Rt. 1, Box 620
Woolwine, VA 24185
540-930-2463
http://igg.com/bdnow/jpi/
JPI makes and distributes the full range of BD Preparations, as well as the BD Compost Starter and the BD Field Starter. Publisher of Applied Biodynamics, *a quarterly newsletter.*

The Pfeiffer Center
c/o Threefold Education Foundation
200 Hungry Hollow Road
Chestnut Ridge, NY 10977
914-352-5020
914-352-5071 fax
http://www.pfeiffercenter.org/
Educational seminars and training on biodynamic farming and gardening.

Demeter Association, Inc.
Britt Rd.
Aurora, NY 13026
315-364-5617
315-364-5224 fax
Contact: Anne Mendenhall
demeter@baldcom.net
Demeter certification for biodynamic produce; publisher of The Voice of Demeter.

Michael Fields Agricultural Institute
W2493 County Rd. ES
East Troy, WI 53120
414-642-3303
414-642-4028 fax
http://www.steinercollege.org/anthrop/mfai.html
Research and extension in biodynamics, publications, conferences, and short courses.
http://attra.ncat.org/attra-pub/biodynamic.html

Permaculture

(http://www.permacultureactivist.net/). http://www.permacultureinter national.org/index.htm

Apples

Phillips, Michael. 1998 *The apple grower.* White River Junction, VT: Chelsea Green Publishing Co.

Outstanding discussions of pests, marketing, and organic philosophy. Oriented to the Northeast.

Index

About the Author and Illustrator

Brian Caldwell is the Farm Education Coordinator for the New York Chapter of the Northeast Organic Farming Association (NOFA-NY). For twenty years, he and his wife, Twinkle Griggs, ran Hemlock Grove Farm, a small commercial organic vegetable and fruit farm in hilly West Danby, New York. They helped found the Finger Lakes Organic Growers Coop and were active members in the Ithaca Farmers' Market (another farmer-owned cooperative) during that time.

He received an MS in floriculture and ornamental horticulture from Cornell University in 1986. As a Cornell Cooperative Extension Educator, Brian advised commercial vegetable and berry growers in a five-county region from 1995 until starting with NOFA-NY in 2002. He keeps from getting too stale by managing a certified organic apple orchard in partnership with West Haven Farm of Ithaca, New York.

Illustrator Jocelyn Langer is an artist, music teacher, and organic gardener, and the illustrator of the NOFA organic farming handbooks series. She illustrates and does graphic design work for alternative media and political events as well as organic-farming-related publications. Jocelyn lives in central Massachusetts.

The special farmer-reviewer for this manual was Jim Crawford; the scientific reviewers were Ruth Hazzard and Rob Wick.